AF552851

POPULAR STUDIES IN PALAEONTOLOGY

OLDEST KNOWN BIRD (ARCHÆOPTERYX) AND A SMALL DINOSAUR (COMPSOGNATHUS).
JURASSIC PERIOD.

ATE XIII. [*Frontispiece.*

POPULAR STUDIES IN PALAEONTOLOGY

By

REV. H.N. HUTCHINSON
B.A., F.G.S

DISCOVERY PUBLISHING HOUSE PVT. LTD.
NEW DELHI-110 002

Published by:
Tilak Wasan

DISCOVERY PUBLISHING HOUSE PVT. LTD.
4831/24, Ansari Road, Prahlad Street
Darya Ganj, New Delhi-110002 (India)
Phone: +91-11-23279245, 43764432
Fax: +91-11-23253475
E-mail: parul.wasan@gmail.com
discoverypublishinghouse@gmail.com
info@discoverypublishinggroup.com
web: www.discoverypublishinggroup.com

Reprinted: **2011**
ISBN: 81-7141-217-3

Popular Studies in Palaeontology

Printed at:
Mehra Offset Press
Delhi

PREFACE

BY SIR W. H. FLOWER, K.C.B., F.R.S.,

DIRECTOR OF THE NATURAL HISTORY MUSEUM.

One of the most important results of the recent progress of science, and one which it is very desirable that we should have fully impressed upon us, is that the living world which we see around us bears an exceedingly small proportion to the whole series of animal and vegetable forms which have inhabited our planet in past ages.

The whole of the knowledge contained in this book is practically the outcome of the scientific work of the present century; by far the greater part of it belongs to the second half of the century. It is difficult now to realize that at its commencement, although it is true that "fossils" had for some time aroused the attention of a few curious persons, and given rise to various and singular speculations about their origin and signification, the only creatures really known, not only to the mass of educated mankind, but even to most naturalists, were those of our own days. The creation of the genus *Palæotherium* (the "ancient wild beast" *par excellence*) by Cuvier in the year 1804, when, by a comparison of its bones, found in the celebrated gypsum quarries of Montmartre, near Paris, with those of all known species of recent animals, he demonstrated for the first time to the satisfaction of the whole scientific world, that vertebrated animals inhabited the earth other than those now found upon its surface, was a great epoch in science. By this demonstration he laid the foundation of the study of the past history of this great group of animals—a study which has developed in this comparatively short period of time to such a marvellous extent, and which has still before it a future of unbounded promise. The most striking illustration of this fact is shown in the wonderful revelations of former animal life, consequent upon very recent explorations carried on in what we commonly call the New World. Numbers of strange forms, entirely unknown or undreamt of before, have been brought to light within the last thirty years, and made known to us by the indefatigable labours of the distinguished palæontologists of the United States, whose names occur so frequently in the following pages. The Old World has also continued to yield up its treasures. Asia, and especially South Africa, are contributing marvellous forms of ancient life, while the most recently explored portion of the earth's surface, Australia, is found to be teeming with evidence that creatures lived there in other days, far more gigantic in proportion and grotesque in figure than any of their modern representatives.

It is well that these results should be collected and placed in an accessible and popular form, as has been done by Mr. Hutchinson in this book. In the restoration of the external appearance of extinct animals, known only by bones and teeth, there is much of imagination, much indeed of mere guesswork, and I should therefore be sorry to guarantee the accuracy of any of the representations of animals in this book, the majority of which were never seen in the flesh by the eyes of mortal man. I think, however, I may safely say that Mr. Hutchinson and his accomplished artist, Mr. Smit, have done their work carefully and conscientiously, and given us in most cases a fair idea of the appearance of the creatures they have endeavoured to depict, according to the best evidence at present available. Of most of these we shall never know more, although it is possible that a different interpretation placed upon some still obscure indications may cause us to modify our views as to their appearance.

The disjointed and often fragmentary bones by which these animals are usually represented in museums convey no ideas whatever to the majority of those who see them. It is quite otherwise with such representations as the figures of this work; and if in some cases the idea conveyed may not be strictly correct, it offers a fair approximation, and at all events gives a vivid conception of some remarkable creature, which in its main outlines cannot be far from the actual reality.

It has been a great satisfaction to me to see, throughout the work, the frequent references to the great collection of remains of extinct animals contained in the Natural History branch of the British Museum, now so admirably arranged by Dr. Henry Woodward and his able assistants, and I trust that a perusal of it will lead to a greater appreciation and closer study of the treasures contained in our national collection.

March, 1894.

AUTHOR'S PREFACE.

For thousands of years men have dwelt upon the earth without even suspecting that it was a mighty tomb of animated races that once flourished upon it as the living tribes do now. Only in very recent times, which men still remember, was the discovery made that the earth has had a vast antiquity; that it has teemed with life for countless ages; and that generations of the most gigantic and extraordinary creatures lived through long geological periods, and were succeeded by other kinds of creatures equally colossal and equally strange. Huge fishes, enormous birds, monstrous reptiles, and ponderous uncouth mammals had possession of a world which as yet knew not man. The vestiges of these creatures are still found in the rocks, their fossil skeletons have been exhumed, and in the light of modern Science their structures and probable habits have been determined.

So numerous and so teeming with interest are the tribes of extinct animals, that the writer found it impossible, within the limits of his previous book, *Extinct Monsters*, to do more than make a somewhat arbitrary selection. Whole orders—and even classes—were entirely left out. From the first we had no intention of dealing with those lower forms of life, so frequently met with by the geologist, which are destitute of any backbone (invertebrates); but, even when confining our attention to the backboned, or vertebrate classes, we found more orders than could be dealt with within the limits of a single volume.

In the present work the writer has attempted to fill up, to some extent at least, the gaps in the last one, and so to present to the reader a simple account (shorn, as far as possible, of the long names and technical terms which repel the general reader) of a certain number of the world's "lost creations." Some of these, if not quite so huge as the "monsters" already introduced to our readers, are yet as wonderful—in some cases, we venture to think, even more so.

It may well be doubted if any more strange and interesting forms of ancient life have ever been brought to light than those very ancient reptiles of the Anomodont Order, described in chap. iv.; which, though reptiles, yet exhibit in their anatomy a strange foreshadowing of features that belong properly to the Mammals —or, perhaps we should say, were once thought to belong only to that class.

We begin with an account of fossil footmarks, a subject which has not been dealt with at any length in the text-books, much less in popular works. There is a fascination in the contemplation of these "footprints on the sands of Time" which appeals strongly to one's imagination; and it is satisfactory to note how more recent discoveries have led to the interpretation of certain facts which were a great puzzle to earlier writers, such as Professor Hitchcock and Dr. Deane, whose beautiful figures of the famous Connecticut Sandstone footprints are of great value.

Part of chap. x. is given to the horse and its ancestors—a subject of universal interest, not only on account of man's fondness for horses, but because of the wonderful series of fossil horses brought to light by geologists, and so clearly interpreted by Huxley, Marsh, and other authorities. These remains constitute the most complete chain of Evolution yet known to the palæontologist. When people say, "Where are those missing links you so often speak of?" he can point with satisfaction to the bones of ancient horses, and show the gradual steps by which the little five-toed ancestor of the Eocene period gradually lost some of its toes, and took on other features until its descendants evolved into the noble animal of the present day.

One cannot help sometimes wishing that a larger number of the many strange and old-fashioned types of life revealed by the Science of Geology in these latter days had continued their existence into the human period, and been added to the number of man's contemporaries on earth. But, judging from the ruthless manner in which so many flowers of creation, whether animal or vegetable, are now being, or have already been, ruthlessly exterminated by the hand of man, one is inclined to think that such a wish ought not to be encouraged. And here two reflections may be allowed: in the first place, there are doubtless very good reasons why some branches of the tree of life have withered and dropped off—although the causes at work seem at present only partly known to geologists and naturalists; and, secondly, supposing that some tribes of Dinosaurs, or uncouth Mammals, had lived on to the present day, what would probably be their fate? Would they be more highly appreciated than those now around us? We may call to mind Mr. Ruskin's remark about grouse-moors and angels; and the sad certainty forces itself upon us that if a modern sportsman were by chance to meet with a Pterodactyl on his moor, his first impulse would be to "shoot it"! Indeed, this foreboding is as it were put into our mouth by the words of a reviewer of our previous work, who, speaking of the creatures therein described, and thinking what big game they might have furnished, remarked, "We only wish we had had the shooting of them!"

It is, perhaps, some consolation to the geologist to reflect that Nature has denied to the nineteenth-century sportsman, with his deadly rifle, the grim satisfaction of shooting her children of old times. Their mortal remains repose safely in the rocks beneath our feet—precious spoil for the geologist's hammer and pick-axe, not for the sportsman's gun!

Some of our former readers have humorously remarked that

a few Dinosaurs and such-like creatures might have constituted a valuable and highly interesting addition to a modern wild-beast show. But here, again, we find it difficult to sympathise, and prefer to think of Dinosaurs as restored by the hand of Mr. Smit and living in the eye of imagination rather than in a common menagerie, to be teased and prodded by the vulgar crowd. They have lived their day, and played their part; let them rest in peace. Doubtless some enterprising showman, catering for the amusement of an idle public, might have been tempted to advertise in big posters "A Jumping Dinosaur," to follow after "The Boxing Kangaroo" or the "Fighting Lion;" but we are glad to think that the Dinosaur has been spared a fate so ignoble, and, instead of this, affords to the biologist a splendid subject on which to exercise his imagination.

As a nation, we spend millions a year in pursuing live foxes; could we not spare a few thousands for hunting up extinct Monsters?

It is to be hoped that the time is not far distant when the Government of this country will no longer trust in its present careless way to private enterprise and liberality for the furthering of scientific discovery, but will give generous grants for the purpose. A good many years ago Mr. Ruskin made some forcible remarks on this subject, which deserve to be widely read.[1] These are his words: "I say we have despised science. 'What!' you exclaim, 'are we not foremost in all discovery? and is not the whole world giddy by reason, or unreason, of our inventions?' Yes; but do you suppose that is national work? That work is all done in spite of the nation, by private people's zeal and money. We are glad enough, indeed, to make our profit of science; we snap up anything in the way of a scientific bone that has meat on it, eagerly enough; but if the scientific man comes for a bone or crust to us, that is another story. What have we publicly done for science? We are obliged to know

[1] *Se*   *nd Lilies*, p. 48, large edit.

what o'clock it is, for the safety of our ships, and therefore we pay for an observatory; and we allow ourselves, in the person of our Parliament, to be annually tormented into doing something, in a slovenly way, for the British Museum, sullenly apprehending that to be a place for keeping stuffed birds in, to amuse our children. If anybody will pay for their own telescope, and resolve another nebula, we cackle over the discernment as if it were our own; if one in ten thousand of our hunting squires suddenly perceives that the earth was indeed made to be something else than a portion for foxes, and burrows in it himself, and tells us where the gold is and where the coals, we understand that there is some use in that, and very properly knight him; but is the accident of his having found out how to employ himself usefully any credit to us? (The negation of such discovery among his brother squires may perhaps be some discredit to us, if we would consider it.) But if you doubt these generalities, here is one fact for us all to meditate upon, illustrative of our love for science. Two years ago there was a collection of the fossils of Solenhofen to be sold in Bavaria, the best in existence, containing many specimens unique for perfectness, and one unique as an example of a species (a whole kingdom of unknown living creatures being announced by that fossil). This collection, of which the mere market worth, among private buyers, would probably have been some thousand or or twelve hundred pounds, was offered to the English nation for seven hundred; but we would not give seven hundred, and the whole series would have been in the Munich Museum at this moment, if Professor Owen had not, with loss of his own time, and patient tormenting of the British public in the person of its representatives, got leave to give four hundred pounds at once, and himself become answerable for the other three! which the said public will doubtless pay him eventually, but sulkily, and caring nothing about the matter all the while, only always ready to cackle if any credit comes of it."

Since these forcible words were written Professor Owen and the leading scientific men of the day have succeeded in inducing a Government to build the Natural History Museum, for which we ought to be very grateful. But although the Museum has been finished for more than ten years, the Treasury still are mean enough to refuse to light up the galleries, either by gas or electricity! The result is, that those who go there to study on foggy days are obliged to turn back, as we have found to our cost. The central hall, however, is lighted; but what is that? Now that the British Museum reading-room is beautifully lighted up by some hundreds of incandescent lamps, let us hope that the Treasury will see to it that our beautiful Natural History collection is not left in darkness whenever London fog prevails, and numbers of citizens turned away in disgust. Another inconvenience, due to the same cause, is that during the winter months (or part of them) the Museum is closed at four o'clock, so that a good day's work becomes impossible.

The author gratefully acknowledges the kind help he has received from palæontologists, both at home and abroad,—help which has lightened his somewhat arduous task, making it easier and pleasanter. He has also received encouraging letters with regard to his previous book, both from palæontologists and general readers, and it is gratifying to find that many who were only acquainted with living animals have been interested in hearing of the results of modern palæontology.

Professor O. C. Marsh, of Yale University, Conn., U.S., has very kindly read many of the proof-sheets, and made valuable suggestions and corrections, especially in the chapters dealing with the more recently discovered Dinosaurs, which he himself has so carefully studied and described. The author's thanks are also due to Professor Marsh for sending a large number of his valuable papers, published in the *American Journal of Science*, and elsewhere.

To Professor E. D. Cope, of Philadelphia, the author is also greatly indebted for advice and help, as well as for many valuable books and scientific papers.

Sir William H. Flower, K.C.B., F.R.S., Director of the Natural History Museum, besides writing a preface, has very kindly read the book before it went to press; for all of which the author is truly grateful.

Again we have to thank Dr. Henry Woodward, F.R.S., Keeper of Geology at the Natural History Museum, for the kind interest he has taken in our work, and for various valuable suggestions.

Mr. A. Smith Woodward, of the Natural History Museum, has again been kind enough to read our proof-sheets and offer valuable advice, especially with regard to fossil fishes (and the restorations of them in Plate II.)—a subject on which his opinion is of great value.[1]

Among other gentlemen connected with the Natural History Museum, to whom the author is grateful for help in various ways, are Mr. C. W. Andrews, Mr. B. B. Woodward, and Mr. F. A. Bather. The attendants and officials of the various galleries and libraries have also always been most courteous and obliging.

The author's thanks are also due to other gentlemen not connected with the Museum, such as Professor H. G. Seeley; Mr. E. T. Newton, palæontologist to the Geological Survey; and Mr. C. Davies Sherborn.

Our good friend, Mr. J. Smit, has once more thrown himself heartily into the task of making more "restorations," and we feel sure that the beautiful drawings he has made for this book will be appreciated by all.

[1] Readers who are acquainted with geological text-books will probably notice that the representations of Old Red Sandstone fishes in Plate II. do not entirely agree with the drawings they have been accustomed to. This is due to the fact that we have derived from Mr. Smith Woodward the latest information with regard to these fishes, and that, in some details, the figures in even the best books are often wrong.

Yours always truly,

Richd. Owen

TABLE OF CONTENTS.

CHAPTER IX.

CHAPTER X.

CHAPTER XI.

CHAPTER XII.

APPENDICES.

LIST OF FULL-PAGE ILLUSTRATIONS.

LIST OF ILLUSTRATIONS IN TEXT.

CHAPTER I.

FOOTPRINTS ON THE SANDS OF TIME.

"They are fraught with strange meanings, these footprints of Connecticut."
—Hugh Miller.

There is a great deal of truth in the saying of Emerson, that "everything in Nature is engaged in writing its own history." The more one studies the changes taking place every day on the surface of the earth, in order to read the riddle of the rocks beneath our feet, which contain Mother Earth's records of her past history, the more one is impressed with the truth of this saying. In fact, it is not too much to say that the whole science of Geology is founded on this idea. The geologist is he who interprets to his fellows the stony documents contained in Nature's "Record Office," and he finds the key to the interpretation of her hieroglyphics in watching her daily actions at the present time. One branch of this science, viz. Physical Geology, deals with the earth's physical features, interpreting in the light of this leading principle their history, and telling us how the river carved out its valley; how the volcano was built up; how the mighty mountains were raised up from the beds of ancient seas, to be carved out by the agents of denudation into all their varied and wonderful features.

Another branch of geology, namely Palæontology, with which we are about to deal in the present work, tells us of the long-lost tribes of plants and animals which, ages and ages ago, found a home on the earth. To many minds this branch of geology, which is simply the natural history of the past, is the most fascinating, with which view we fully sympathise. Putting aside the study of fossil plants (which is a small branch of the subject), we may say of Palæontology, that it interprets to us the world's "lost creations." In this branch of geology the records are not so much the rocks themselves as the fossil bones they so often contain.

In some cases, as we shall presently show, the rocks contain additional evidences of very considerable value, as throwing light on the habits of the creatures, or on their natural surroundings; but bones, shells, and other hard parts of animals, are the foundation on which the science of Palæontology is founded. And here, again, we find the same principle at work—viz. that the past must be read in the light of the present. However many ages ago it was, whether millions or billions of years ago, that these primæval inhabitants of the world enjoyed their existence, the same unbroken laws of nature—the visible expressions of a Divine and All-powerful Will—were at work, fulfilling His purposes, as now. Flesh and blood were then what they are now, and fulfilled the same functions. Bones grew then as they grow nowadays. To those bones were attached muscles which expanded and contracted just as muscles do now. Wings were used for flying, fins and paddles for swimming, legs for walking, teeth for masticating food, just as they are now. In fact these primitive inhabitants of the antique world, however different in bodily shape from those we see around us now, lived under the same universal laws of Physiology as we ourselves do.

Palæontology, then, is the science which, in the light of Comparative Anatomy and Physiology, rehabilitates the world's ancient inhabitants, clothing their dry bones with flesh, and enabling us in imagination to see them as they were when they walked this earth. It will be our endeavour in the present work to present to our readers a certain number of these antique animals—birds, beasts, and fishes, whose mortal remains have been buried up and preserved, often with singular completeness, in the rocks of the earth's crust.

But, although fossil bones and skeletons are the chief material at our command for this purpose, yet the series of stratified rocks contains here and there other kinds of evidence, valuable in their way, such as foot-marks, tracks, burrows, coprolites, or droppings, and even ripple-marks and the impressions of rain-drops. It is with these evidences that we propose to deal in the present chapter. Let us see what can be learned from such humble and apparently insignificant records, of some of the creatures that once trod this earth.

The intelligent observer who has strolled along the strand of the seashore at low water, must have often seen the surface of the exposed sands deeply rippled by the waves of the ebbing tide, and have noticed the trails of molluscs, and the meandering furrows and ridges produced by worms, or annelides, and the tracks of crabs, and sometimes the footprints of birds and of dogs or other quadrupeds, that have walked over sand or mud while it was yet plastic and sufficiently firm to retain the markings impressed upon it. Under certain conditions these apparently evanescent characters are indelibly fixed on the stratum, and in rocks of immense antiquity successive layers of sandstone and shale, through a thickness of many hundred feet, are found deeply furrowed with the ripples of the waves that flowed over them, pitted by the rain that has fallen upon them, and impressed with the footmarks of bipeds and quadrupeds that traversed the sands whilst the surface was in a moist and yielding state. Even on some of the most ancient of rocks, such as those of the Cambrian system, jelly-fish have left indelible impressions of their soft round bodies!

Speaking of the wonderfully enduring nature of certain impressions known to geologists, the sagacious Dean Buckland said, in an address to the Geological Society: "The historian or the antiquary may have traversed the fields of ancient or of modern battles, and may have pursued the line of march of triumphant conquerors, whose armies trampled down the most mighty kingdoms of the world. The winds and storms have utterly obliterated the ephemeral impressions in their course. Not a track remains of a single foot or a single hoof of all the countless millions of men and beasts whose progress spread desolation over the earth. But the reptiles that crawled upon the half-finished [1] surface of our infant planet have left memorials of their passage, enduring and indelible. Centuries and thousands of years have rolled away, between the time in which these footsteps were impressed by tortoises upon the sands of their native Scotland and the hour when they were again laid bare and exposed to our curious and admiring eyes. Yet we behold them stamped upon the rock, distinct as the track of the passing animal upon the recent snow; as if to show that thousands of years are but as nothing amidst eternity, and, as it were, in mockery of the fleeting perishable course of the mightiest potentates among mankind." [2]

Every form of animal life that, writhing, crawling, walking, running, hopping, or leaping, could leave a track, depression, or footprint behind it, might thereby leave similar lasting evidence of its existence and also, to some extent, of its nature. The interpretation of such evidences of ancient life has exercised the sagacity of naturalists since Dr. Duncan, in 1828, first inferred the existence of tortoises in certain sandstones in Dumfriesshire from the impressions left on them. The vast number and variety of such impressions has raised up a distinct branch of Palæontology, to which the name Ichnology [3] has been given.

[1] This expression is a survival from the teaching in vogue fifty years ago. The world was not in an unfinished state during the period of the New Red Sandstone.

[2] *Bridgewater Treatise*, vol. i. p. 251.

[3] Greek—*ichnos*, footstep; *logos*, discourse.

We will now give a brief account of the results which have been arrived at in this branch of inquiry. To begin with one of the lowest forms of animal life—the worms. The class Annelida comprises the so-called ringed worms, including the leeches and earth-worms, and the sea-worms. As might have been expected, earth-worms are unknown in the geological record; for their soft bodies were not likely to be preserved even in the most favourable kinds of deposits. But, in some cases, the hard jaws of marine worms have escaped destruction. Fossil worm-jaws are abundantly found in some parts of the Cambrian, Silurian, and Carboniferous systems (see Appendix I.). The so-called "Conodonts" are believed by many authorities to be the jaws of worms; and such remains are also found in strata of the Mesozoic and Tertiary eras. Besides these rather mysterious little bodies, which some have taken to be the teeth of primitive fishes—such as our modern Hag-fish,—a good many worm-like markings are found in muddy and sandy sediments all through the stratified series. In many cases the true nature of these remains is still a matter of doubt. The visitor to the Natural History Museum at South Kensington will find in Gallery No. 11 a very fine and large collection of fossil tracks and footmarks of all kinds (Wall-case No. 7). Some of these are probably vegetable remains; but others are certainly the tracks of molluscs or of crustaceans. Long burrows of marine worms occur plentifully in some rocks of Cambrian and Silurian age, and have been figured under the names *Scolithus*, *Histioderma*, and *Arenicolites*. They are nearly straight, and descend vertically through the rock. Such are abundant in that ancient formation the Potsdam Sandstone of North America; in the Clinton formation, also of that country; and in the Stiper-stones of Shropshire. Even in the Pre-Cambrian rocks of the west of Sutherlandshire there have been discovered of late years some long dark lines which are believed to be the burrows of marine worms pulled out to great lengths by the "shearing," or pulling-out process to which these rocks have been long ago subjected. For the full and complete interpretation of many of the curious markings known to geologists, a more accurate knowledge of the markings made by living animals will doubtless be necessary.

Various worms of the present day, such as the common lug-worm, are known to form long, wandering, tortuous channels in the sand of the seashore, a little distance below the surface. These worms feed on particles of organic matter scattered through the sand or mud, through which it eats its way. Such burrows cross and intersect each other in various ways, and as the worm proceeds on its course, they become filled up in the rear by the sand which has passed through its body. This is how worm-casts seen on the seashore at the present day are made. It appears, in the light of more recent researches, that many markings found in some of the more ancient (Palæozoic) rocks, and which have been formerly described as "fucoids," *i.e.* sea-weeds, under such names as *Palæochorda*, are in reality the filled-up burrows of marine worms. These have now been re-christened *Planolites*. But there are some who consider them to be tracks made on the surface, *not* burrows.

A great many true trails, or tracks of worms, etc., that is markings made by the animal dragging its soft body over the surface of wet sand or mud, are found in the stratified rocks. But, in the present state of knowledge, it is very difficult to distinguish between those formed by worms and others made by molluscs or even crustaceans. The fossil known as *Nereites*, from Silurian slates at Wurtzbach, is probably the track of a worm. On the other hand, some of the tracks attributed to worms may have been really made by gastropod molluscs, such as whelks. One fossil track, known as *Crossopodia*, resembles the track made by a living *Purpura lapillus*, a well-known sea-shell. It is only in those strata which are very favourable to the preservation of organic remains that we can expect to find any trace or impression of the actual body of such a frail and perishable thing as a worm; but, incredible as it may seem, fossils of this nature occur in that most wonderful formation—the famous Solenhofen limestone, in which so many valuable treasures have been found; also in the Eocene slates of Monte Bolca (Italy). In these rare cases the form of the worm's body is actually seen, and the fossilised jaws occur in their natural position. Examples of these interesting specimens are beautifully rendered by chromolithography in Professor Zittel's monumental work *Palæontographica*.

Several geologists, such as Poulett-Scrope, Strickland, Buckland, Salter, and others, have published the results of careful comparisons of tracks made by living animals on the sands of the seashore, or on flat surfaces of mud left exposed by the drying up of a pond, or by other causes, but have not given drawings of the recent markings on which their conclusions are based. Professor Emmonds and Professor T. McKenny Hughes, however, have figured some recent tracks in illustration of fossil ones. The former geologist came to the conclusion that certain imprints upon some very old rocks—the Taconic Shales of Maine and New York States—were made by the soft fragile larvæ of insects which existed at an early period in the world's history.

We now pass on to the consideration of certain impressions in the old Potsdam Sandstone of North America, which have been most carefully studied by the late Sir R. Owen. In the year 1851, Logan exhibited before the Geological Society of London a small slab of sandstone showing some footprints, and a plaster cast from a longer surface of similar description. The original, weighing upwards of a ton, is in the Museum at Montreal connected with the Geological Survey of Canada. The locality where it was found is on the left bank of the River St. Louis, at the village of Beauharnois, on the south side of the St. Lawrence, about twenty miles above the city of Montreal. Owen, in his first paper, came to the conclusion that the tracks were those of a tortoise. But further research caused him to alter this opinion. We only mention this to show much care is required even on the part of the best naturalists to read the meanings of fossil tracks. A portion of the impressions now under consideration is shown in Fig. 1. They consist of a series of well-defined impressions continued in regular succession for four feet, and more; but only clearly for four feet. In this four feet there are thirty successive groups of footprints on each side of a furrow. The number of prints is not the same in each group. Where they are best marked,

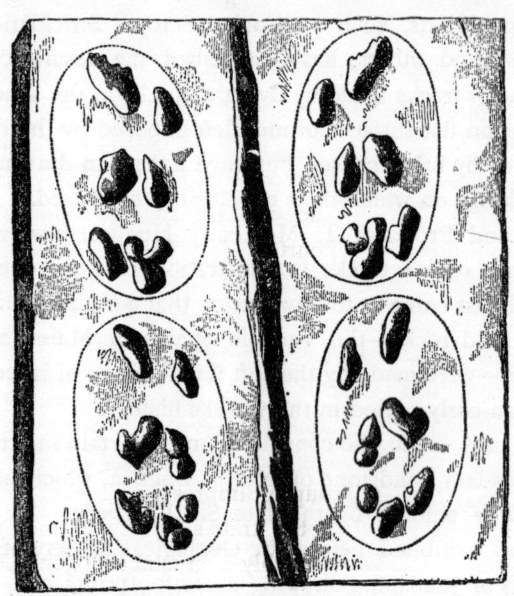

FIG. 1.—Tracks (*Protichnites*), probably of a crustacean, from Potsdam Sandstone, North America. (After Owen.)

as in our figure, we see three prints in one group, two in the next and two in the third, followed by a repetition of the three prints (in our illustration each of the three groups is enclosed in an oval). These three groups (of 3-2-2 impressions) are distinctly repeated in succession along the whole series of tracks on both sides of the furrow. It will be noticed that in each pair of impressions the innermost pair are of equal size, but of the outer ones each is a little bigger than the last. An important point is that there are no mark of toes or nails. Their edges are not sharply defined, but are rounded off, and there is a slight variation in the form and depth of the corresponding impressions on each side of the furrow. But the reader will see from the figure that they do correspond with each other. Thus, take the three tracks at the bottom of the lower group on the right side of the furrow; the innermost of this group of three may easily be identified with the innermost track of the group of three on the left side of the furrow. And so with the two groups of three belonging to the two upper sets of impressions, each of which is enclosed in an oval.

These very ancient tracks are known to geologists under the name *Protichnites*, and the creature that produced them must have made no less than fourteen impressions, seven on the right and seven on the left, each time its legs were set to work. There seems to be doubt of this, because the groups of tracks, as marked out in our illustration, occur again and again in successive series so similarly and so regularly as to admit of no doubt that they were made by repeated applications of the legs, and these must have been capable of being moved so far in advance as to keep clear of the previous group of impressions. Sir Richard Owen concludes his account of the tracks by saying that the creature which made them was probably a crustacean genus, and that it may have had three pairs of limbs employed in locomotion, each of which was split up into two or more parts so as to make in walking either two or three tracks. The shape of the pits so clearly seen on these slabs of the old Potsdam Sandstone (although they have been rubbed and polished by the action of glacier-ice) suggests that they were made by the hard and partly pointed, partly blunt, terminations of the limb of a crustacean, such as a crab or lobster. But this creature moved directly forwards, not like a crab, but like a lobster or a king-crab. The furrow that runs between the tracks was probably made by a tail.

The question then arises—what sort of a crustacean was it that made these tracks? Great caution is required in dealing with a problem of this kind, as will be seen from the following words of Sir R. Owen: "In all probability no living form of animal bears such a resemblance to that which the Potsdam footprints indicate as to afford an exact illustration of the shape and number of the instruments, and the mode of locomotion of the *Protichnites.*" The imagination is baffled in the attempt to realize the extent of time past since the period when the creature was in existence which moved upon the sandy shores of the ancient Cambrian period, to which the "Potsdam Sandstone" belongs.

In about the year 1830, much interest was excited by the discovery of footmarks, resembling those of land tortoises, on the exposed surfaces of slabs of sandstone of Triassic age, in a quarry at Corncockle Muir in Dumfriesshire, of which an interesting account was published by the Rev. Dr. Duncan. Regular tracks, indicating the slow progression of a small four-footed animal over the surface while the stone was in the state of moist sand, were traced on the blocks of sandstone, when separated by the quarry-men, along the lines of their stratification. In one instance there were found twenty-four consecutive impressions, forming a track with six distinct repetitions of the marks of each foot, the front feet differing from the hind feet. The appearance of five claws was discernible on the impressions of each fore paw. In 1853, Sir William Jardine published a splendid folio work in which he fully described these footprints; it was illustrated by full-sized lithographs coloured after Nature.[1]

The footprints occur in the Dumfriesshire Sandstones, in different patches, in several localities, but are best seen either where naturally exposed in the valleys of the Esk, the Nith, and the Annan, or in the quarries in those districts where they are worked for building material. One of those areas, of considerable extent, fills up the bottom of nearly all the upper basin of the

[1] *Ichnology of Annandale* (Edinburgh, 1853; folio).

Annan Valley above the ridge at Dormont Rocks. The beds are about two hundred feet thick, and present even surfaces. It is a curious fact, observed by the author of the above-mentioned book, that all the footprints are impressed as if the animal had walked from west to east. As a rule the creature seems to have walked in a straight line, but sometimes the tracks turn and wind in different directions. The paces are generally even and uninterrupted, seldom diverging much aside, showing little stoppage for food, or for a scuffle with a neighbour, which sometimes accompanied them. They appear more as the tracks of animals passing at once across some tide-receded estuary, in pursuit of some well-known and favourite grounds which were periodically sought after for some particular purpose. But it must be borne in mind that the impressions figured in this important work are not all similar in shape, and were probably due to different animals. They often show the effects of a peculiar pushing-back motion, which may be noticed in living tortoises. Dean Buckland, who was interested in these impressions, caused a living tortoise to walk on soft sand, clay, and paste, and found a fairly close correspondence between the tracks thus made and those of Corncockle Muir.

In 1831, Mr. Poulett-Scrope, an English geologist, described some small tracks made by a crustacean on a rock of the Jurassic period, known as the Forest Marble.

Some years previous to 1856, a series of strange impressions was found in a quarry in the lowest part of the Millstone Grit formation at Rhodes Wood, near Tintwhistle, Cheshire. The proprietor, Mr. Rhodes, was much struck with them, from the fact that they bore a resemblance to the marks of a human foot. The workmen also were struck with the resemblance, and, when they first showed them the impressions, remarked, "Master, some one has been here before us!" For several weeks the quarry was visited by many hundreds of people from Glossop and the surrounding neighbourhood. The common opinion was that the tracks were the footprints of some of Noah's family! This strange idea seems to have been founded on another equally strange, viz. that the Ark had rested on some neighbouring hills. But to return to the tracks; the distance between the impressions was two feet ten inches, and several of the impressions were thirteen inches long. Mr. Waterhouse Hawkins saw them, and thought that they resembled the supposed chelonian track figured by Jardine. Mr. Binney's conclusion (who read a paper on these tracks before the Geological Society) was that they were made on wet sand by a heavy slow-moving animal, like a tortoise, with irregular gait.

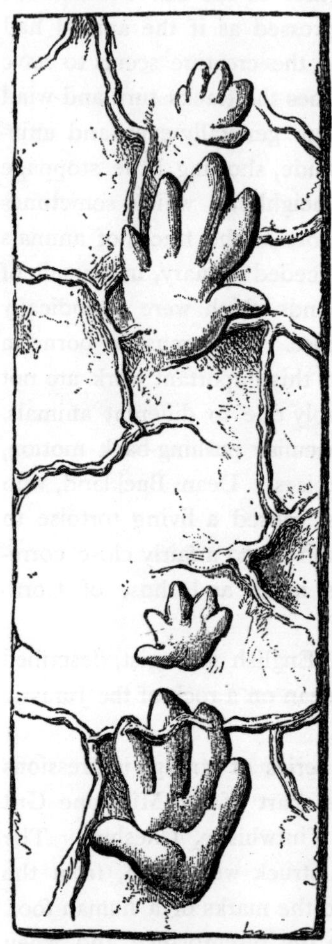

FIG. 2.—Footprints of *Cheirotherium*, in the Bunter Sandstone, Hessberg, near Hildburghausen.

Soon after the discovery of footprints at Corncockle Muir, another discovery was made in strata of the same geological age at Hessberg, near Hildburghausen, in Saxony. These footprints, however, were evidently made by somewhat large creatures, in which the fore paws were much smaller than the hind ones (see Fig. 2). Subsequently, similar tracks were observed on slabs of Triassic Sandstone in the quarries at Storton, near Liverpool. Others, again, have been found in Cheshire. The Museums at Warwick,

Warrington, and Liverpool, are rich in impressed slabs from the New Red Sandstone. Numerous fine specimens may be seen in the Museum of the Geological Society, the Museum of Practical Geology, in Jermyn Street, and in the Natural History Museum, Cromwell Road, which is a part of the British Museum, Gallery No. 11. Some of the slabs from Storton are covered with small round pits, or depressions, produced by rain-drops that fell while the surface was soft and impressible. The footprints from Storton are seen on the face of each successive stratum of sandstone, the corresponding surface of the overlying stone presenting, in relief, casts of the imprints and other markings. The hollow impressions of the feet are always on the upper surfaces of the slabs, and the convex casts on the under side of each layer or stratum, the latter fitting closely into the former. In our illustration of tracks from Saxony only *one* pair of tracks is given; there would, of course, be another pair parallel with them. The double lines in between are casts of sun-cracks formed as the mud dried in the sun.

These "footprints on the sands of time" follow one another in pairs—one small, the other large, each pair being in the same line, and some fourteen inches in advance of the other. Each footmark has five toes, and the first, or great toe, is bent inward like a thumb, and is alternately on the right and left side of both the large and small footprints, which, except in size, resemble each other. The German geologist who first described them in 1834, Dr. Kaup, proposed the name of *Cheirotherium*[1] for the great unknown animal that left the footprints, in consequence of the resemblance, both of the fore and hind feet, to the impression of a human hand. No *certain* remains of the creatures whose tracks we are now considering have yet been discovered in the same strata. But in these rocks and others of the same geological age in England and Germany there have been obtained skulls, teeth, and bones of amphibians,

[1] Greek—*cheir*, hand; *therion*, beast.

known as *Labyrinthodonts*, of which we shall have more to say in Chapter III. Some of the salamander-like amphibians of the Triassic period, we now know from later discoveries, attained to a considerable size.

As Sir R. Owen points out, the impressions of the *Cheirotherium* resemble the footprints of a modern salamander in having the short outer toe of the hind foot projecting at right angles to the line of the middle toe, but yet are not identical with those of any known batrachian or reptile. Still it has been conjectured by the same great authority, as well as by others, that these footprints were the work of the creatures now known as Labyrinthodonts, which have left their remains in rocks of the Carboniferous, the Triassic, and the Permian ages. He argued as follows: (1) There is proof from the skeleton that the *Labyrinthodon* had hind limbs larger than its fore limbs. (2) That the size of the known species of *Labyrinthodon* corresponds with the size of the footprints of the *Cheirotherium*. (3) The *Labyrinthodon* occurs in the Triassic strata, in which the *Cheirotherium* impressions are found. (4) That no remains of animals likely to have produced such impressions are found in these strata except the *Labyrinthodon*.

Our friend, Professor W. J. Sollas, has described some interesting footprints from South Wales, which probably were made by a Dinosaur of the Triassic age.[1] A friend of his was, in 1878, passing through the village of Newton Nottage, in Glamorganshire, when his attention was arrested by some three-toed footprints on a slab of rock, deeply impressed and rendered particularly visible by the slanting rays of the setting sun. Casts of them were afterwards made by the curator of the Cardiff Museum (Mr. J. Storrie).

To show how valuable geological finds are often neglected through ignorance of their real worth, it may be mentioned here that this slab is even now lying in a corner of the village green, in

[1] *Quarterly Journal of Geological Society*, xxxv. (1879), p. 510.

front of the church; formerly it lay in front of the steps of the inn, where it consequently suffered more or less wear. The impressions remind one of some of those described by Professor Hitchcock (see p. 19), and resemble more especially those belonging to his genus *Brontozoum.* Other specimens were afterwards found in the same locality. Professor Sollas had casts made of impressions of the feet of a living emu in the Clifton Zoological Gardens, for the sake of comparison, and found a good deal of agreement between the two. Nevertheless, from what we now know of Dinosaurs, it would be unwise to say that the impressions were made by birds (see p. 21).

Professor W. C. Williamson has described some very interesting impressions from Cheshire. They were found, by a former pupil of his (Mr. J. W. Kirkham), near Weston Point. They are unlike those of the *Cheirotherium*, previously described, and differ from all others yet found in showing very distinctly what are probably the *marks of scales.* The form of the foot also differs in being more quadrate. Professor Williamson says it reminds him of certain footprints found by Dr. King in the Carboniferous rocks of Pennsylvania (see p. 63). The arrangement of the scales corresponds closely with that seen on the foot of a modern alligator. The impressions suggest a saurian much more than an amphibian.

These impressions figured by Professor Williamson remind one a good deal of some tracks described by Professor Huxley, in his paper on a "New Red Sandstone Crocodile from the Triassic Strata of Elgin,"[1] which may have been made by that ancient leviathan.

We now pass on to give some account of those famous footprints in the Connecticut Valley, of which probably all geologists have heard. The River Connecticut, in part of its course through the country which bears its name, and in the northern district of

[1] *Quarterly Journal of Geological Society*, xv. (1859), p. 440.

the adjoining State of Massachusetts, flows through a valley of sandstone of the Triassic age. Successive layers of this rock are exposed all along considerable tracts of country. From this circumstance, and from the facility of transport afforded by the river, numerous quarries have, for many years, been worked in various parts of the valley, near the water's edge. The many footprints contained in these rocks were observed much earlier than the date (1828) in which the Rev. Dr. Duncan first described the tracks at Corncockle Muir (see p. 10). They have been very fully described and figured by Professor Ed. Hitchcock [1] and Dr. J. Deane.[2]

As far as Dr. Deane could learn, the first specimen was ploughed up in South Hadley, in 1802, by a boy. This specimen is now in the Appleton Ichnological Cabinet.[3] So strikingly did the tracks resemble those of birds, that they were familiarly spoken of as the tracks of poultry, or of "Noah's raven."

It was not until the year 1836 that any attempt was made to describe the tracks scientifically. The year previous some flagging stones were obtained in Montague for the streets of Greenfield, by a Mr. Wilson, who observed impressions upon them, which he regarded as tracks of "the turkey tribe." These were observed by Dr. Deane, who sent casts of them to Professor Hitchcock. Professor Hitchcock gave his first account of them in *The American Journal of Science*, in 1836. He propounded the idea that they were the tracks of birds—a view which was not adopted by scientific men at the time, though afterwards many came

[1] Professor Ed. Hitchcock, *Ichnology of New England*. (Boston, 1858.)

[2] Dr. J. Deane, *Ichnographs from the Sandstone of Connecticut River*. (Boston, 1861.)

[3] The late Honourable Samuel Appleton, of Boston, left by his will a large sum of money to be appropriated by the trustees under his will to benevolent and scientific purposes. Those trustees accordingly appropriated ten thousand dollars to the erection of the Appleton Cabinet at Amherst, a museum, of which the lower story is entirely devoted to fossil footmarks. A still larger collection is now in the Museum of Yale University, but only a part is on view.

round to his opinion. We have perused his work with great interest, and cannot but admire the care with which he studied the tracks and endeavoured to interpret their meaning, although his conclusions now require a good deal of modification.

Professor Hitchcock chose to give fanciful names to the creatures that made the tracks, such as *Brontozoum giganteum*, "the huge animal giant;" *Polemarchus gigas*, "the huge leader in war," and so on. He concludes a description of some of these creatures in the following words: "Such was the fauna of the sandstone days in the Connecticut Valley. What a wonderful menagerie! Who would believe that such a register lay buried in the strata? To open the leaves, to unroll the papyrus, has been an intensely interesting though difficult task, having all the excitement and marvellous developments of romance. And yet the volume is only partly read. Many a new page, I fancy, will yet be opened, and many a new key obtained to the hieroglyphic record. I am thankful that I have been allowed to see so much by prying between the folded leaves. At first, men supposed that the strange and gigantic races which I described were mere creatures of imagination, like the Gorgons and Chimæras of the ancient poets. But now that hundreds of their footprints, as fresh and distinct as of yesterday, impressed upon the mud, arrest the attention of the sceptic on the ample slabs of our cabinets, he might as reasonably doubt his own corporeal existence as that of these enormous and peculiar races."

Professor Hitchcock's work on this subject was much appreciated by his countrymen, as the following facts will show. It was resolved by the Commonwealth of Massachusetts, in the year 1857, that his *Geological Report on the Sandstone of the Connecticut Valley*[1] be printed, and one copy furnished to each member of the executive and legislative departments of the Government of that year, and one copy to each town and city in the Commonwealth. Not content with this, it was resolved in the following year

[1] *Ichnology of New England.*

that more copies be printed, that one hundred be given to the author, and others for the purpose of international exchange. Englishmen cannot but admire this good example, and would doubtless be glad if it could be followed at home, where so little is done for the cause of Science by our rulers.

But, in justice to Dr. Deane, we must not omit to mention his part in the description and figuring of these interesting records. Unfortunately he died before the completion of his beautifully illustrated work on the subject. He drew with his own hand a large number of lithographs from the actual slabs; and these drawings will always be of the greatest use to students of Ichnology. He worked at the subject in his spare moments, during an active professional career, and with great enthusiasm, as the following extract from his unfinished work will show.

"An indescribable interest is imparted by opening the long-sealed volume that contains the records of these extinct animals. The slabs were uncovered and raised under my supervision, and page after page, with their inscriptions, revealed living truths. There were the characters, fresh as upon the morning when they were impressed, reminding the spectator of the brevity of human antiquity, and of the frail tenure of human works! On that morning, how long ago no one can tell, or ever will know, gentle showers watered the earth, an ocean was unruffled, and upon its borders primæval beings enjoyed their existence and inscribed their eventful history."

And now to sum up the results of their work. Most of the tracks so fully described and figured by these two authors were probably made by amphibians and reptiles, and it is doubtful if any of them were made, as they supposed, by birds. They vary extremely, both in size and character. While some are only half an inch long; others, like the huge *Otozoum*, of Hitchcock (see Fig. 3, left hand corner), are twenty inches long, and show a stride of three feet! Some of the creatures that made the tracks had five toes, some four, and many of them only three. Again, some had hind feet and fore feet of nearly equal size, and evidently walked or crawled on all-fours.

We have here reproduced, from Professor Hitchcock's work, a highly interesting plate (Fig. 4), which shows that some of the animals had large hind feet and small fore feet, and that they sometimes put down the latter, as kangaroos do, for there are the impres-

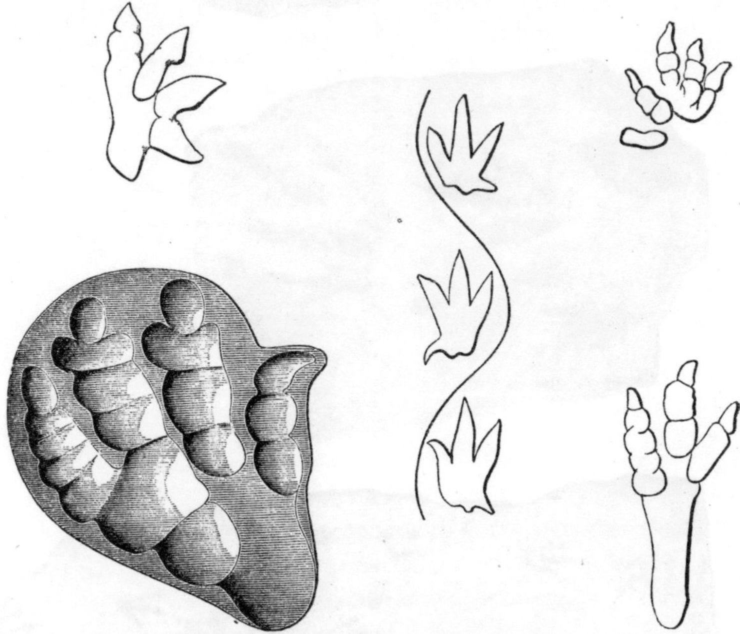

FIG. 3.—Footprints from Connecticut Sandstone. The three tracks in the middle show the mark left by the tail, probably of a Dinosaur. (After Hitchcock.)

sions of them. However, such impressions are rare, and, as a rule, the creatures, many of which were probably Dinosaurs, walked on their hind limbs, thus producing those three-toed bird-like impressions which are shown on the slabs represented in our figure. It is hardly likely that they hopped on their hind legs as kangaroos

FIG. 4.—Footprints of Dinosaurs, from Connecticut Sandstone, showing impressions of the small fore feet (1 and 3). (After Hitchcock.)

do; they were probably too heavy for that; but they may have made something like a hop in bringing up their hind feet, as we have seen the kangaroos do in the Zoological Gardens. Another very interesting point brought to light by the researches of Hitchcock and Deane is, that on some slabs may be seen what is believed to be the impression of the tail of a Dinosaur as it dragged along over the moist sand of the seashore or estuary where these antediluvian creatures of old disported themselves (see Fig. 3).

Those impressions which Hitchcock believed to have been made by birds show a regular increase in the number of joints of the toes; the inner toe having two, the middle one three, and the outer one four joints. Now, it happens, by a remarkable coincidence, that in the case of birds the inner toes have three, the middle toes four, and the outer toes five joints, but the last two joints in each case make but one division of the track, so that their tracks correspond with those we are considering. The discovery that some Dinosaurs have but three toes that were used (functional toes), and that they had the same number of joints in their toes as birds, has naturally led scientific men to the opinion that probably *all* the tracks above described were made, not by birds, but by Dinosaurs. It may be pointed out in defence of this opinion, that, although the two classes of Birds and Reptiles are *now* widely separated, yet in certain former geological periods there was no such gulf as now divides them. At the time when the Connecticut Sandstone tracks were made on the shores of a narrow inland sea, there must have been in existence animals which, if we saw them now, might sorely puzzle us to decide whether to call them reptilian birds, or bird-like reptiles.

During the Jura-Trias period there was in the region of the Connecticut Valley a shallow sea, connected by a narrow outlet with the ocean. Into this the tides flowed and again ebbed, leaving extensive flats of mud or sand, ribbed with ripple-marks. A passing shower pitted the soft mud, and the sun, coming out again from the breaking clouds, dried and cracked it. Our Dinosaurs and other creatures sauntered or ran near the margin of the shore. The tide came in again, carrying with it fine sediments, gently covered the tracks, and preserved them for ever. This occurred constantly for many ages, about the time when the Triassic period came to a close.

In the year 1882, reports were published of the discovery of large footprints—supposed to be human—in a certain sandstone, near Carson, Nevada, U.S., of which a brief account was given in our former work.[1] These are probably the tracks made by a big extinct sloth of the Pleistocene period. The wonderful series of footprints of reptiles, birds, and mammals discovered by M. Desnoyers, in certain Eocene strata near Paris, are reserved for description on p. 184.

[1] *Extinct Monsters*, p. 185 (2nd edit.).

CHAPTER II.

FISHES OF BYGONE DAYS.

"In the endeavour to complete the natural history of any class of animals, the mind seeks to penetrate the mystery of its origin, and, by tracing its mutations in time past, to comprehend more clearly its actual condition, and gain an insight into its probable destiny in time to come."—SIR RICHARD OWEN.

SWEDENBORG, in his writings, describes how the angels examined a dead man in order to read the story of his life. One might have supposed that they asked him questions; but no, they simply examined his body, and from that read out to him all the things that he had done. They unrolled his fingers, for whatever acts these performed had left their record on them. So with his brain; whatever thoughts had been his were found to have left a record. There is much truth in this striking idea; and Mr. Ruskin has said the same of human faces. But leaving the realm of psychology, and turning to the solid earth, we find countless numbers of dead bodies of animals sealed up in the rocks beneath our feet, which may be looked upon as a great graveyard. The flesh, of course, is gone; but the bones which supported that flesh are there, and the work of the palæontologist is to take up these bones, and, as far as is possible to finite human intelligence, read therefrom the histories of the countless tribes of creatures that flourished on the earth long ages ago. The method is precisely similar in the two cases. Every bone has its meaning, and every skeleton can be made (in the hands of competent anatomists) to tell its story.

Since the majority of strata deposited in times past on the earth's surface were formed in seas, lakes, and estuaries of rivers, it might naturally be expected that the fossilized remains of fishes would be frequently met with; and such is the case. This class of backboned creatures is well represented in most of the geological museums.

Even within the last twenty years or so, palæontologists have learned a great deal more about fossil fishes; but knowledge in this department has only kept pace with the general progress of palæontology, which has been wonderfully rapid, as the following facts serve to show. In the year 1830, Dr. Samuel Woodward published a *Table of British Organic Remains*, and in this book (the first of its kind) two pages sufficed for all the British fossil vertebrates! Then came Professor J. Morris's well-known *Catalogue of British Fossils* with thirty pages for vertebrates; and now our friends, Messrs. Smith Woodward and C. D. Sherborne, have published a most valuable *Catalogue of British Fossil Vertebrates*, of 396 pages, with nearly 200 pages devoted to fishes only.

Every fossil-collector is delighted when he comes across a fossil fish; one gets rather tired of everlasting mollusca, and a fish cropping up now and then is a great encouragement to a geologist. The finest collection of fossil fishes in the world is that in our Natural History Museum, under the care of Mr. Smith Woodward. The large room devoted to this purpose contains 450 genera, and 1250 species, the results of years of labour on the part of many celebrated collectors. Of living species there are about 10,000. The collection has of late been enriched by the purchase of two splendid collections—one that of the Earl of Enniskillen, F.R.S., from Florence Court, Enniskillen, Ireland; the other that of Sir Philip de Malpas Grey-Egerton, F.R.S. (Trustee of the British Museum). Here also are to be found the collections of Dr. Gideon Mantell and Mr. Frederick Dixon. With these, and many other valuable donations and purchases, the national collection now contains most of the British types figured by Agassiz in his famous works, *Researches on Fossil Fishes* and *Monograph of the Fossil Fishes of the Old Red Sandstone.* Any one visiting this collection will be much struck with the wonderfully perfect state of preservation of many of the specimens, such as those from the famous Solenhofen Limestone, and others obtained from the Limestone of Hakel and Sahel-el-Alma in the Lebanon, through the energetic labours of the Rev. Prof. E. R. Lewis, F.G.S., late of the Syrian Protestant College, Beirût (see p. 47).

As we intend, in this chapter, to deal chiefly with Old Red Sandstone fishes, it may not be out of place to mention at least some of the collectors from these rocks whose labours have been of great service. The place of honour must certainly be given to the late Hugh Miller. The well-known researches of this distinguished geologist brought to light an abundant fish fauna in the Old Red Sandstone, a formation which was previously thought to be barren ground for the geologist in search of fossils. " Half my closet walls," he says, " are covered with the peculiar fossils of the Lower Old Red Sandstone; and certainly a stranger assemblage of forms has rarely been grouped together—creatures whose very type is lost, fantastic and uncouth, and which puzzle the naturalist to assign them even their class; boat-like animals, furnished with oars and a rudder; fish, plated over, like the tortoise above and below, with a strong armour of bone; other fish less equivocal in their form, but with the membranes of their fins covered with scales; creatures bristling over with thorns; others glistening in an enamelled coat, as if beautifully japanned; the tail, in every instance, among the less equivocal shapes, formed not equally, as in (most) existing fish, on each side the central vertebral column, but chiefly on its lower side—the column sending out its diminished vertebræ to the extreme termination of the fin. All the forms testify of a remote antiquity—of a period 'whose fashions have passed away.' The figures on a Chinese vase or an Egyptian obelisk are scarce more unlike what now exists in nature than are the fossils of the Lower Old Red Sandstone."

Robert Dick of Thurso was also a worker in the Old Red Sandstone; and the combined collections of these two geologists are now in the Edinburgh Museum of Science and Art. The late Lady Gordon Cumming, of Altyre, made extensive excavations at Lethen Bar, Nairnshire, providing many specimens for description by Agassiz, and forming the collection now in the Forres Museum. "Her object in making this collection," says Hugh Miller, "was the illustration of the geology of the district, and all she sought on her own behalf was congenial employment for a singularly elegant and comprehensive mind." The Lower Cornstones of Hereford, Worcester, and Monmouthshire have long been noted for specimens of certain peculiar forms of fish (to be presently described), such as *Cephalaspis* and *Pteraspis.* Numerous specimens may be seen in the Museums of Hereford, Ludlow, Worcester, and Oxford. Many of these were collected by the Rev. T. T. Lewis, of Aymestry, and Dr. Lloyd, of Ludlow, who submitted them to Agassiz through Sir R. Murchison. The late Rev. W. S. Symonds, of Pendock, was also an enthusiastic collector, and by his numerous writings[1] set others to work in the same direction. Monmouthshire was also vigorously explored by the late J. E. Lee, F.G.S., who left all his specimens to the national collection. The late Dr. D. M. McCullough, of Abergavenny, obtained a large series of the shields of *Pteraspis* (see p. 42). The Rev. John Anderson, D.D. (author of *Dura Den*), has explored the fish-bearing strata of Dura Den, Fifeshire.

As the reader is probably aware, fishes form the lowest class of backboned or vertebrate creatures. Their skin is covered with numerous scales, composed of a dense durable substance, which, in some families, is strengthened by the addition of bony plates, like those we see on the body of a sturgeon. But in whatever way their bodies are covered, they are thus provided with a strong

[1] *Records of the Rocks*, etc.

yet flexible coat of armour, affording suitable protection to beings peculiarly exposed to external injuries, from the nature of the regions they inhabit, and perhaps also the state of warfare with each other in which they are constantly engaged.

In dealing with the more ancient and now extinct genera of fishes, geologists have often to be satisfied with very imperfect evidence; sometimes only a few scales, or a spine from the back, or a few teeth. The teeth of certain ancient sharks, on account of their hard enamelled surface, are met with in great abundance in some strata; in the Silurian and Carboniferous rocks the large and formidable spines of these ferocious fishes are also abundant (see Fig. 6), while others of the Ganoid order are well preserved on account of their coat of armour, consisting of hard bony plates with a surface of enamel, like that of the Bony Pike of North America. In chalk strata the cup-like vertebræ, or joints of the backbone, are frequently met with, as well as sharp-pointed teeth. Deposits of fine mud or of calcareous ooze (such as afterwards became consolidated into limestone) seem to have been very favourable for the preservation of ancient fishes, especially ganoids with their hard coats of mail. Hence we often find in the soft clays and marls of the Tertiary rocks, in the chalk of England, in the fine lithographic limestone of Bavaria, or in the limestones of the Lebanon, fishes very perfectly preserved; and not only individuals, but groups, with the scales, fins, head, teeth, and even the capsule of the eye, all undisturbed from their original position. And, better still, there are formations in which, as we shall show presently, the palæontologist comes across not only the bones, scales, and teeth, but, to his great delight, finds a perfect outline marked on the stone, showing the true shape of the creature; in these exceptional cases it may indeed be said that Nature has left us lithographs of ancient fishes.

Some of the specimens from the Jura limestone show a wonderful state of preservation. Sir R. Murchison had in his own collection a small slab of marl from Aix, in Provence, containing scores of small fishes as perfect as if recently embedded in soft mud. In the chalk, many of the fishes have not been squeezed flat by pressure from overlying strata, the body being almost as perfect in form as if the original had been surrounded by soft plaster of Paris before sinking into its grave of grey ooze on the floor of the old chalk sea.

A curious discovery was made by a late Lord Greenock, who found between the laminæ of a block of coal, from the neighbourhood of Edinburgh, a mass of petrified intestines, distended with coprolite (fossil excrement), and surrounded with the scales of a fish which Professor Agassiz referred to the great *Megalichthys* (see p. 45). Some of the wormlike bodies so abundant in the Solenhofen limestone (see p. 7) would appear to be either the petrified intestines of fishes or the contents of the same, still retaining the form of the twisting tube in which they were lodged. To these remarkable fossils the name *Cololithes* has been given.

Naturalists and palæontologists have been, and still are, much in doubt as to the best and most natural way in which to classify or arrange fishes, both fossil and recent forms. What characters should be taken as a basis for classification? At one time it was the fashion to take for this purpose the nature of the skeleton (whether "bony" or "cartilaginous"), as well as the number and position of the fins. The first point is an important one, and is still used for this purpose; but the fins do not form so important a character. Later on, Agassiz, one of the greatest authorities on fishes, unfortunately proposed a classification based entirely on the nature of the scales; and for a long time his scheme was accepted by both geologists and naturalists. But this plan has been abandoned, because it was found to be misleading, and separated groups which should come near to each other. By his scheme fishes were arranged in four groups, or orders, as follows:—

(1) *Placoid.* Ex.: the shark. (2) *Ganoid* (from the brilliant

surface of the enamel). Ex.: gar-pike (*Lepidosteus*). (3) *Ctenoid.* Ex.: the common perch. (4) *Cycloid.* Ex.: the salmon.

The term "Ganoid" is still in use, and the order retained, but not the others.

Taking the skeleton alone, we find that fishes fall into two great divisions, viz. bony fishes and cartilaginous fishes. The essential difference in the skeleton of these two groups consists in the presence or absence of earthy matter (phosphate and carbonate of lime) in the materials of which they are constituted. A herring, or a salmon, is an example of a "bony" fish. In the second group the skeleton is transparent, being composed chiefly of cartilage or gristle, as in the sharks. But some fishes are intermediate in this respect, and their skeletons contain a certain proportion of lime, though a small one.

In some genera, certain portions of the skeleton, as the bodies or centra of the vertebræ, are cartilaginous, while other parts, such as the spines and the ribs, are osseous.

The best authorities of the present day, Dr. Traquair, Dr. Günther, Mr. Smith Woodward, Professor Cope and others, are not agreed as to the precise way in which fossil and living fishes should be classified; but the following scheme is one of the most useful that have been put forward.

Leaving out altogether the Lancelet (*Amphioxus*), which has long been considered to be the lowest form of fish, but which differs from them in so many respects that it is better to refer it to a separate class altogether, we have six orders, thus—

Order I. *Cyclostomata* (Lampreys and Hags). Their skeleton is cartilaginous, and the skull is not separated from the vertebral column. There are no fins, and the body is eel-like. They have a round mouth, with teeth arranged in a circular manner; hence the name Cyclostomata, or "circle-mouthed."

Order II. *Plagiostomi* (Sharks and Rays). So called because the mouth is placed transversely on the under side of the body. In these fishes the skeleton remains cartilaginous throughout life.

The skull is well developed, but consists of a single cartilaginous box, there being no distinct bones. The back bone, or vertebral column, is sometimes composed of distinct vertebræ (of which the bodies, or centra, may be partly turned to bone), but it may be quite simple (notochordal). In most fishes of this order the extremity of the tail is slightly turned upwards, and the lower lobe of the tail-fin is much larger than the upper, producing what is known as a "heterocercal" tail (see next page). They belong to the "Placoid" order of Agassiz, because the scales take the form of more or less numerous detached, grains (or sometimes plates), tubercles, or spines, scattered here and there in the skin. When very small and close set, as in the "dog-fish," such a skin is called "shagreen." But in the case of the rays these ossifications take a singular shape, viz. a disk from the upper surface of which springs a sharp curved spine.

Order III. *Chimæroidei* (Chimæras). These fishes resemble the sharks in many important features. Their skeleton is wholly cartilaginous, and they have a notochord instead of a true backbone; the vertebræ being merely represented by slender rings. Ex.: *Edaphodon* (Cretaceous).

Order IV. *Dipnoi* (or Mud-fishes). The skeleton in this order is notochordal. They breathe by means of air-sacks answering to lungs, as well as by gills, like other fishes. In consequence, they are able to quit the water at times, and live in a torpid condition in the mud. The forelimb is very different to the pectoral fin of other fishes, and seems to foreshadow the forelimb of amphibia and reptiles. Ex.: *Lepidosiren* from the Amazons, and *Ceratodus* from Queensland.

Order V. *Ganoidei.* These fishes are remarkable for their highly enamelled scales and head-shields, and are represented by hosts of extinct forms, but at the present time they are confined to a few lakes and rivers. Ex.: *Polypterus* of the Nile, and *Lepidosteus*, or Gar-pike, of North America. Their skeletons may be either cartilaginous, or ossified, or both.

Dr. Günther, of the Natural History Museum, is in favour of bringing together the second, third, fourth, and fifth orders, and making one sub-class of them, viz. the *Palæichthyes*, or Ancient Fishes; but these differ so much among themselves that his proposal has not been accepted by Traquair and others.

The tails of fishes, living and extinct, present a feature of considerable interest. In the greater number of existing species the backbone terminates in a triangular plate of bone, to which the tail-fin is attached symmetrically; and its figure is either rounded or divided into two equal lobes, or branches. These tails are termed "homocercal," *i.e.* "even-tailed," as in the case of a salmon. But in the ganoid and placoid fishes we find the vertebral column rising up into the upper lobe of the tail; the caudal fin appearing like a rudder, and the lower lobe being feeble in proportion, as in the shark. This form is called "heterocercal," *i.e.* "unequal-tailed."

Order VI. *Teleostei* (or Bony Fishes). These are believed to represent the highest stage of development to which fishes as a class have attained. Now, it appears to be established that, to some extent at least, the development of an individual animal is, as it were, an epitome of that of the race from which it springs. Here we have an example illustrating this highly interesting law; for, if a bony fish be examined before its development has been completed, *i.e.* while it is yet an embryo, the tail may be seen to be first pointed (as in some more primitive forms of the present day), then turned up, as in the heterocercal type, and lastly it becomes symmetrical as in the highest types (*Teleostean*).

Perhaps no fact is more clearly recorded on those stony tablets we call strata than that from time to time in the world's history (as read by geologists from Nature's record) higher types of animals have appeared on the scene. In other words, we find plainly written in the rocks a *Law of Progress.* This important conclusion may safely be accepted; but, having accepted it, let us see what follows therefrom. The fish, with its backbone, is, of course, a higher type than the mollusc or the insect. It seems to follow of necessity, then, that there must have been a time in the world's history when this type did not exist. It may be difficult for us to conceive of a world tenanted only by coral polyps, medusæ, crabs and lobsters, worms, insects, and suchlike humble creatures, with no fish in the sea, no reptiles, birds, or mammals; and not only does the doctrine of Evolution plainly teach this, but such a conclusion is distinctly favoured by the evidence derived from a study of fossils and their distribution in time. The earliest fossil-bearing strata, known as the Cambrian series, although they have been carefully explored, have at present yielded no evidence of the existence of fishes at that time.

The advent of back-boned animals was a great event in the world's history. Endless possibilities loomed in the future for these children of life, in which, as Mrs. Fisher (Miss Buckley) says, "the solid skeleton should be—not a burden for the soft body to carry, as in the sea-urchins, snails, insects, and crabs—but an actual support to the whole creature, growing with it and forming a framework for all its different parts."[1] And not only so, but the skeleton of a vertebrate animal is no mere mass of hardened dead matter; it is a living and growing framework which goes on increasing in size and strength as fast as its possessor grows. Again, the spinal chord, or vital thread of the animal, which in other creatures has no special protection against shocks or jars, is here protected by being encased in the hard bony joints, or vertebræ, of the backbone. Little need we wonder, then, that creatures so favourably endowed for the daily struggle should have been able to strike out in so many directions, producing such various and successful types of organization as the fish, the reptile, the bird, and the mammal,—all, as it were, new branches of the great tree of life.

[1] *Winners in Life's Race*, p. 7. One of the best popular books on Natural History ever written.

No one can yet say what was the primitive form from which all the vertebrates were derived; this is a very interesting problem, and one on which anatomists have been much exercised; but, at present, the question must be regarded as still *sub judice.* Whether it was some form of worm, or whether a primæval relative of the arachnid "king crab" (such as the *Pterygotus*, described in our former work)[1] turned itself upside down, rearranging limbs and head; these are questions still admitting of endless discussion, and probably beside the mark, although no doubt suggestive. It is just possible that future geologists may be fortunate enough to find, buried up in some very early stratified rock, the remains of a forerunner of the backboned race; but, unless it possessed some hard parts, the probability is against such a lucky find turning up. Still the problem need not be given up as hopeless; for there is important aid to be derived from two other sources—(1) the study of embryos of living animals; (2) from certain very primitive types, such as the Lancelet, and the Ascidian, or "Sea-squirt," which may be either survivals from very ancient types, or degraded forms which have so far degenerated as to form a kind of borderland between vertebrates and invertebrates, and which cannot be said to possess a true backbone in the ordinary sense of the word. But we must leave all these interesting questions, and pass on to consider fish life as it was in some of the early geological ages. It must always be borne in mind that at best the geological record is an incomplete one, depending as it does on rather accidental circumstances. For example, the lowest order of fishes, represented by Lampreys and Hag-fishes (*Cyclostomata*) at the present day possess no hard parts beyond minute horny teeth, and therefore we can hardly expect to find in the stratified rocks any evidence of their former existence. We cannot hope to get a complete history of any class, even of fishes, preserved in the rocks, although no doubt fishes are better represented than

[1] *Extinct Monsters*, p. 26 (new edit.).

any other vertebrate fossils. Yet there are certain little fossils known as "conodonts," found in Cambrian, Silurian, and Devonian rocks in Russia, which are very suggestive of the tiny teeth of Lampreys and Hag-fishes, although they consist of phosphate of lime instead of horny matter. This interpretation, however, is open to doubt, and some authorities think that the bodies in question really represent the jaws of marine worms, or even of Trilobites! (see p. 5).

Putting aside this somewhat doubtful evidence, the earliest true fossil vertebrate yet known is a single head-shield of *Scaphaspis ludensis*, from Lower Ludlow strata in Shropshire. This fossil belongs to a group presently to be described under the Ganoid order. But a little above this is the famous Ludlow Bone-bed, celebrated for its numerous remains of ancient fishes allied to our Sharks, and belonging to the order Plagiostomi. This highly interesting stratum, first discovered by Sir Roderick Murchison near the town of Ludlow, usually consists of one or two thin layers of brown bony fragments near the junction of the Old Red Sandstone and the Ludlow rocks, and is only a few inches thick. Here we have abundant evidence that shark-like fishes were flourishing vigorously during the Silurian period; but it is impossible to say exactly what they were like, because, from the perishable nature of their framework, they have only left behind them spines, granules of their shagreen, teeth, etc.[1] However, there is every probability that they resembled the Port Jackson

[1] "The fish-bed of the Upper Ludlow rock," says Hugh Miller, "abounds more in osseous remains than an ancient burying-ground. The stratum, over wide areas, seems an almost continuous layer of matted bones, jaws, teeth, spines, scales, palatal plates, and shagreen-like prickles, all massed together, and converted into a substance of so deep and shining a jet colour, that the bed, when first discovered, conveyed the impression, says Mr. Murchison, that it enclosed a triturated heap of black beetles." Coprolites, or lumps of fossil excrement, have also been found containing small but recognisable portions of shell-fish that lived at the same time, thus confirming the idea that the sharks of the period (of which little but dorsal spines, such as *Onchus*, are now left) lived upon shell-fish.

Shark now living in Australian waters, viz. *Cestracion Philippi*, shown in Fig. 5. Although a shark, this genus is a harmless one, and its fortunate survival through long ages of the past to present times throws much light upon the detached teeth so frequently met with in strata of the Carboniferous, Jurassic, and Cretaceous ages. It is provided with a large number of teeth of various shapes and sizes. Those in front are sharp-pointed and well adapted for seizing their prey, while those further back are arranged in oblique rows, and, from their flat surfaces, suitable for crushing the shell-fish on which they feed. Were it not for the useful lesson in Palæontology conveyed by this living fish, there can be little doubt that a good many species would have been founded on

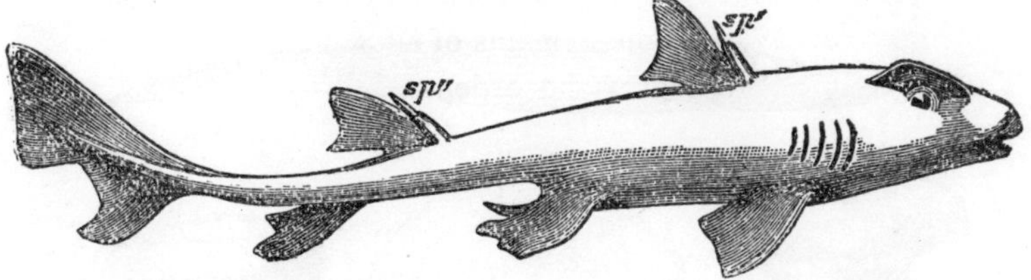

FIG. 5.—The Port Jackson Shark (*Cestracion Philippi*). Australia. *sp'*, *sp''*, spines.

different kinds of teeth belonging really to the same fish. As it is, several species founded on detached teeth may eventually be included in a single one.

The spines of this kind of fish are generally capable of being elevated and depressed, and not only serve the purpose of defence, but, in many instances, afford support and protection to the soft rays of the fin; forming, as it were, a movable mast by which the sail can be spread out or lowered at pleasure. Such spines are very frequently met with, not only in the Ludlow bone-bed but in strata of various ages; they have received the name of "Ichthyodorulites" (fish-spine stones). Fig. 6 shows some spines and teeth of plagiostomous fishes from rocks of the Carboniferous age. Visitors to the Natural History Museum at South Kensington will find in the gallery devoted to fossil fish (No. VI. on plan) a splendid collection of such spines (Wall-case No. 1).

Of all the organisms of the Old Red Sandstone, the *Pterichthys*, or "winged fish," is the most peculiar. It was first introduced to geologists by Mr. Hugh Miller, who, in his delightful work, *The Old Sandstone*, says, "I fain wish I could communicate to the reader the feeling with which I contemplated my first-found

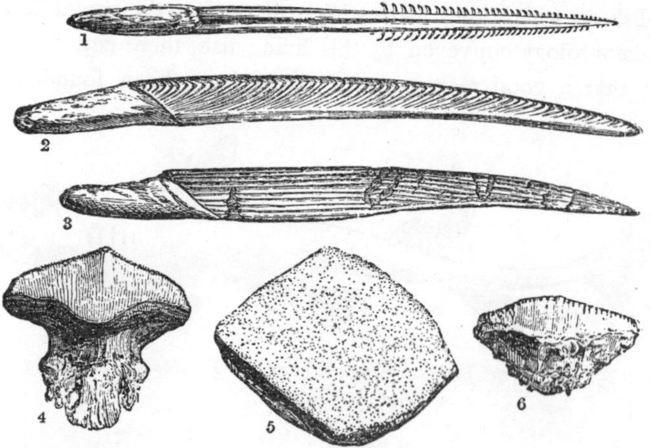

FIG. 6—Spines and teeth of shark-like fossil fishes. 1. Fin-spine of *Pleuracanthus*. 2. Of Gyracanthus. 3. Of *Ctenacanthus*. 4. Tooth of *Petalodus*. 5. Of *Psammodus*. 6. Of *Ctenoptychius*. All from the carboniferous rocks.

specimen. It opened with a single blow of the hammer; and there, on a ground of light-coloured limestone, lay the effigy of a creature fashioned apparently out of jet, with a body covered with plates, two powerful-looking arms articulated at the shoulders, a head as nearly lost in the trunk as that of the ray or sun-fish, and a long angular tail. My first-formed idea regarding it was, that I had discovered a connecting link between the tortoise and the fish. The body much resembles that of a small turtle; and why, I asked, if one formation gives us sauroid fishes, may not another give us chelonian ones? Or if in the Lias we find the body of the lizard mounted on the paddles of the whale, why not find in the Old Red Sandstone the body of the tortoise mounted in a somewhat similar manner?"

Hugh Miller submitted some of his specimens to Sir R. Murchison (then engaged in his famous researches on the Silurian System), who was much interested, and sent them on to the great naturalist Agassiz. Though broken, the specimens were declared by him to represent an ancient fish—a conclusion which was anticipated by Murchison. The latter authority told Hugh Miller he could not conceive of their belonging to a reptile; and if not fishes, they seemed to show more approach to crustaceans than to any other class.

Mr. Miller found in the Lower Old Red Sandstone of Scotland a great number of fish remains, in consequence of which he was convinced that the sea of that period was amazingly fertile in fish life, and as thoroughly occupied with fishes as our friths and estuaries when the herrings congregate most abundantly on our coasts. "There are evidences too sure to be disputed that such must have been the case. I have seen the ichthyolite beds, where washed bare in the line of the strata, as thickly covered with oblong spindle-shaped nodules as I have ever seen a fishing-bank covered with herrings; and have ascertained that every individual nodule had its nucleus of animal matter,—that it was a stone coffin in miniature, holding enclosed its organic mass of bitumen or bone, its winged, or enamelled, or thorn-covered ichthyolite." He was also much impressed with certain evidences which seemed to him to point to the conclusion that over a wide area in the North of Scotland the fish of this period were somehow involved in sudden destruction. These are his words: "The same platform in Orkney, as at Cromarty, is strewed thick with remains which exhibit unequivocally the marks of violent death. the figures are contorted, contracted, curved; the tail in many instances is bent round to the head; the spines stick out; the fins are spread to the full, as in fishes that die in convulsions. The *Pterichthys* shows its arms extended at their stiffest angle, as if prepared for an enemy. The attitudes of all the ichthyolites on this platform are attitudes of fear, anger, and pain. The remains, too, appear to have suffered nothing from the after attacks of predaceous fishes; none such seem to have survived."

It is difficult to assign a cause to such wholesale destruction, but there are ways in which fishes of the present time are sometimes killed in very large numbers. The poisonous gases emitted from submarine volcanoes have often wrought great destruction; again, violent floods from mountain streams coming into the sea laden with mud and fresh water suggest a possible explanation.

The carcases of dead fishes have even helped to consolidate the strata of the Old Red Sandstone. For, just as a plaster cast boiled in oil becomes thereby denser and more durable, so the oily and other matter coming from decomposing fish operated on the surrounding sand or mud so as to make it more compact. The well-known flagstones of Caithness are an instance. They owe their peculiar tenacity and durability to the dead fishes that rotted in their midst while yet they were only soft mud. In no other part of the world, perhaps, can the builder set a large flagstone on its edge with the assurance of it holding together in that position. A large portion of the county of Caithness formed, before its upheaval, what may truly be termed a *piscina mirabilis.*

But, to return to our *Pterichthys,* it was a wonderful little creature, quite different to any living fish. The head and anterior or forward part of the trunk were defended by overlapping ganoid plates, those of the trunk forming a buckler; but the head is sharply separated from the trunk—probably by a movable joint (see Fig. 7). It is also remarkable for the pair of limbs or appendages, near the head, which are hollow, movably articulated, and once jointed. Fig. 8 is a restoration after Pander, but in some respects not correct. Nothing is yet known of the precise nature of the nose, mouth, and jaws. Fig. 7 gives a more correct idea, and is after Dr. Traquair, but partly modified by Mr. Smith Woodward (to whom we are indebted for much kind help and advice in writing this chapter[1]). In the latter figure the two eyes are marked by two circles, and between these is a

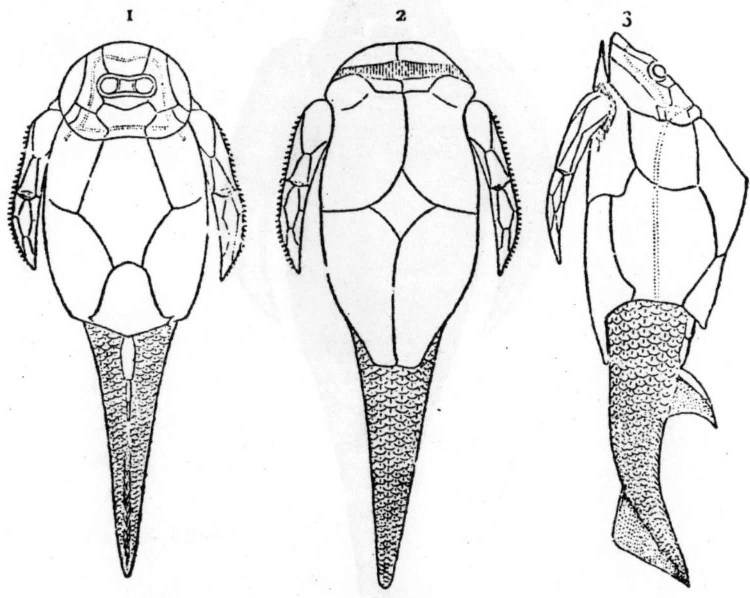

FIG. 7.—*Pterichthys Milleri.* (Partly modified after Traquair.)
1. From above. 2. From below. 3. Side view.

small loose plate which appears to have lodged a small third but rudimentary eye (some of the *Labyrinthodonts* seem to have also possessed a third rudimentary eye, see p. 56). The middle drawing shows the probable nature of the jaw-plates. Behind the armour-plated box of the body comes a tapering, scaly tail, which

[1] See a valuable paper by Mr. Smith Woodward in *Natural Science*, vol. i. p. 596. "The Forerunners of the Backboned Animals."

is often fairly preserved. With regard to the two powerful and jointed limbs, Sir R. Owen thought that they served to aid the fish in shuffling along the muddy or sandy bottom if left dry at low water. The two fins, one on the back and the other on the tail, indicate a certain amount of swimming power, though not any great rapidity. No doubt the swimming was chiefly accom-

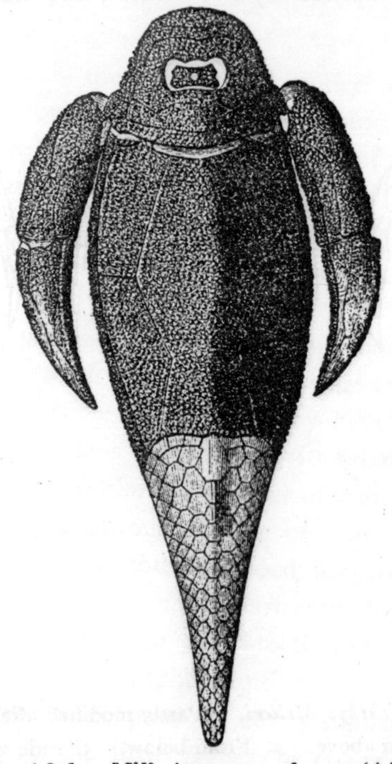

FIG. 8.—*Pterichthys Milleri*, upper surface. (After Pander.)

plished by means of the two broad arms—we can scarcely call them fins—which remind us so much of the two large appendages of the *Eurypterus*, a huge crustacean that lived in the same Old Red Sandstone sea.[1] These extinct crustacea of the order *Merostomata* were then both numerous and powerful; it has

[1] *Extinct Monsters*, p. 30 (new edit.).

therefore been suggested that perhaps the *Pterichthys* and its allies took to mimicking them. "Imitation," they say, "is the sincerest flattery," but in this case it may have been a wise precaution!

When first discovered, our *Pterichthys*, being so strangely unlike other fossil fishes, was the occasion of some curious blunders. Thus the Rev. Dr. Anderson, of Newburgh, figured and described it as a beetle, in the pages of a Glasgow topographical publication—*Fife Illustrated.* Although *Pterichthys* was small, other members of the same group were much larger, while some attained to great dimensions. It is a question not yet decided whether they should be included under the ganoid order, though for present purposes they may be placed there.

The group we are now considering is known by the name Placodermata, on account of the strong and large plates protecting their bodies and heads. A splendid collection of such fishes may be seen in the fossil-fish gallery at South Kensington (Wall-case No. 4).

Coccosteus [1] (see Fig. 9), another genus, varies in length from

FIG. 9.—*Coccosteus* restored.

a few inches to two feet in length. Its tail was destitute of scales, and its back-bone was only partly ossified, no vertebræ having been observed. *Homosteus* (allied to the above genus) was a gigantic form, of which the remains were first found in Russia. Specimens have also been obtained from Scotland and the Eifel.

[1] Greek—*kokkos*, berry; *osteon*, bone, from the tubercles on its bony plates.

Dinichthys[1] is another huge placoderm from Devonian strata in North America, described by Professor Newberry, with a probable length of from fifteen to eighteen feet. Its teeth resemble those of the living African Mud-fish (*Protopterus*).

A specimen of *Titanichthys* recently found was six feet across the head, with orbits on the eyes three inches in diameter.

Dr. Traquair is of opinion that these fishes may turn out to be a curiously modified group of Dipnoi or Mud-fishes, while Professor Cope has suggested that some of them (the Pterichthys and others) are not fishes at all, but an ancient order of "Sea-Squirts" (*Tunicata*)!

Another group is represented by certain contemporaries of *Pterichthys*, such as *Scaphaspis Pteraspis* (see Fig. 10) and others. The second of these, as already mentioned (p. 34), comes from

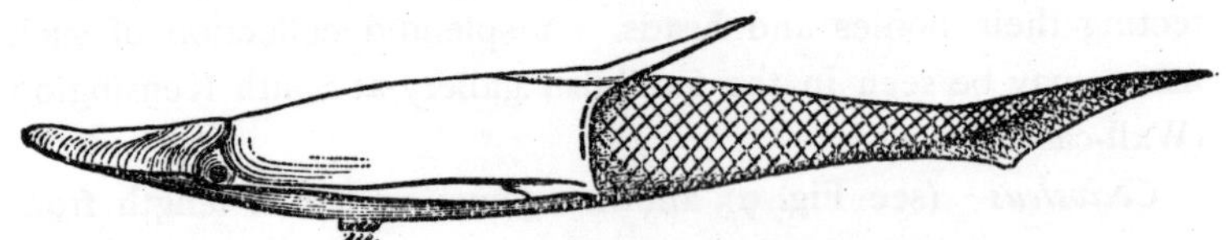

FIG. 10.—An Old Red Sandstone fish, *Pteraspis*. (Restored after Ray Lankester.)

the Lower Ludlow rocks of the Silurian system. Its head is covered with a large buckler, composed of one or more plates; the tail is covered with scales, but no internal skeleton of any kind has been met with. This is the oldest fish at present known to geologists.

Fig. 11 shows the most recent interpretation of *Cephalaspis*,[2] a member of the same group, which was found and graphically described by Hugh Miller. It was not a large fish, most of the specimens in the Natural History Museum (Table-case No. 35) being

[1] Greek—*deinos*, terrible; *ichthys*, fish.

[2] This figure also appears in the paper in *Natural Science*, by A. Smith Woodward, already quoted.

not more than seven or eight inches long. The genus was established by Agassiz. In 1847 Dr. Rudolph Kner published a memoir

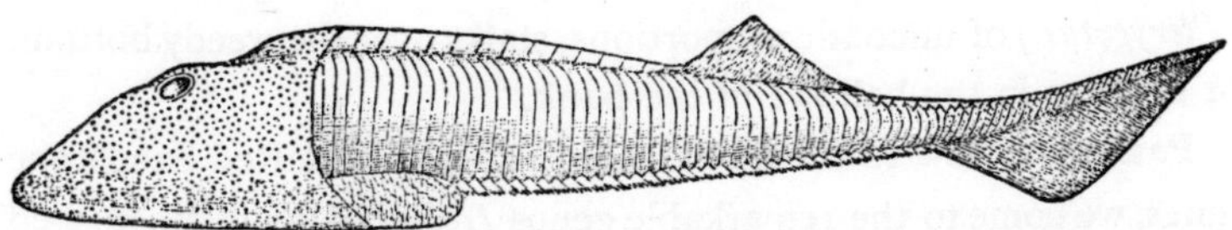

FIG. 11.—Restoration of an Old Red Sandstone fish, *Cephalaspis Murchisoni*, from specimens discovered by G. H. Piper, Esq. (From *British Museum Catalogue of Fossil Fishes.*)

to prove that certain species were the internal shells of Cuttlefishes allied to the *Sepia*. Röemer thought that they were crustacea; and, like *Pterichthys*, they certainly do bear some resemblance to *Pterygotus* and its allies. Later on, in 1856, Professor Huxley published a most detailed and careful account of the structure of the fossil head-shields of the genus (or family), and proved conclusively that *Cephalaspis* was neither a cuttlefish nor a crustacean, but a fish; pointing out, at the same time, a certain resemblance to the living *Loricaria*, a siluroid fish.

Hugh Miller's graphic description of the state of things during the deposition of the Cornstone formation and of its fauna is as follows: "The curtain rises and the scene is new. The myriads of the lower formation have disappeared. We are surrounded, on an upper platform, by the existences of a later creation. There is sea all round, as before; and we find beneath a dark-coloured bottom, thickly covered by a dwarf vegetation. The circumstances differ little from those in which the Ichthyolite beds of the preceding period were deposited; but forms of life essentially different career through the green depths, or creep over the ooze. Shoals of *Cephalaspides*, with their broad arrow-like heads, and their slender angular bodies, feathered with fins, sweep past like clouds of crossbow bolts in the ancient battle. We see the distant gleam of scales, but the forms are indistinct and dim; we can merely ascertain that fins are elevated by spines of various shape and pattern ; that of some the coats glitter with enamel; and that others—the sharks of this ancient period—bristle over with minute thorny points. A huge crustacean (*Pterygotus*) of uncouth proportions, stalks over the weedy bottom, or borrows in the hollows of the bank."

Passing on now to the true Ganoid fishes of Old Red Sandstone times, we come to the remarkable genus *Holoptychius*, distinguished by the peculiar structure of its scales, the enamelled surfaces of which are marked with undulated furrows. The whole body is covered by thick scales of this kind. It was first introduced to the notice of geologists by Professor Flemming, in a paper read before the Wernerian Society, and published by him in 1831 in *Cheek's Edinburgh Journal.* But at that time only detached scales and the fragment of a tooth had been found. These enigmatical remains were afterwards submitted to Agassiz, who referred them to the fish *Gyrolepis*, though not without considerable doubt. Not long after, however, an almost entire specimen of the creature was discovered in the sandstones of Clashbennie, by the Rev. James Noble, of St. Madoes, a gentleman who, by devoting his leisure hours to Geology, did, in his time, much to extend a knowledge of the upper part of the Old Red Sandstone. Agassiz therefore attached his name to the fossil, calling it *Holoptychius Nobilissimus.* This important discovery proved that the scales in question did not belong to *Gyrolepis*, but to the genus just mentioned. The body of this specimen measures a foot across by two feet and a half in length, exclusive of the tail, which is wanting. It is now in the Natural History Museum. Those who examine the specimen for themselves will see that the scales are deeply sculptured, as the name implies.

There is a remarkable family of Ganoid fishes termed "Sauroid" by Agassiz, in consequence of certain peculiarities in their organization which are found in no other animals of this class, but exist in reptiles. Of this family there are but two genera living; namely the *Lepidosteus*, or "Gar-pike," of which there are several species inhabiting the rivers of America; and the *Polypterus*, with two species, one inhabiting the Nile, the other the rivers of Senegal. Their teeth are large, and conical, and striated, as in the crocodile and the extinct *Plesiosaurus*, etc. The skeleton is osseous all through. Even in their soft parts many analogies to reptilian structures are seen.

These two African and American fishes are, for some unaccountable reasons, the only surviving representatives of a tribe of ancient fishes which must have abounded in the seas of Palæozoic and Mesozoic times. Their relics have often been mistaken for those of amphibians; particularly the teeth, which, from their large size, and conical shape, etc., might easily be taken for those of *Labyrinthodonts* (see p. 52). Rhizodus (which partly corresponds to *Megalichthys*, and attained a length of nine feet), is found in Carboniferous rocks.

The *Osteolepis*[1] may be regarded as a good example of the old Ganoid fishes. It was one of the first discovered of the Caithness fishes, and received its name, in the days of Cuvier, from the bony character of its scales, ere it was ascertained that it had numerous contemporaries, and that to all and each of these the same description applied. The scales of the fishes of the Lower Old Red Sandstone were all coated externally with enamel. It requires skill such as that possessed by Agassiz to determine that the uncouth *Coccosteus*, or the equally uncouth *Pterichthys*, with their long tails and tortoise-like plates, were *bonâ fide* fishes; but there is no possibility of mistaking the *Osteolepis:* it is obvious to the least practised eye that it must have been a fish, and a handsome one. *Dipterus*,[2] another Ganoid of the period, is so named from its two dorsal fins, but in this respect it is not peculiar. Another genus, *Diplopterus*,[3] also owes its name to the order and arrangement of its fins, which

[1] Greek—*osteon*, bone; *lepis*, scale.
[2] Greek—*dis*, twice; *pteron*, wing.
[3] Greek—*diploös*, double; *pteron*, wing.

were placed fronting each other and in pairs. It had a large head and a wide jaw.

"In their most striking characteristics, however, the three genera seem to have agreed. In all alike, scales of bone glisten with enamel; their jaws, enamel without and bone within, bristle thick with sharp-pointed teeth; closely jointed plates, burnished like ancient helmets, cover their heads, and seem to have formed a kind of outer table to skulls, externally of bone and internally of cartilage; their gill-covers consist each of a single piece, like the gill-cover of a sturgeon; their tails were formed chiefly on the lower side of their bodies; and the rays of their fins, enamelled like their plates and their scales, stand up over the connecting membrane, like the steel or brass in that peculiar armour of the Middle Ages whose multitudinous pieces of metal were fastened together on a groundwork of cloth or of leather. All their scales, plates, and rays present a similar style of ornament. The shining and polished enamel is mottled with thickly set punctures, or rather punctulated markings; so that a scale or plate, when viewed through a microscope, reminds one of the cover of a saddle."[1]

In Plate II., author and artist have endeavoured to give a glimpse of fish life as it was in one of the inland seas, or lakes, of the Old Red Sandstone period. To the right is seen the *Holoptychius*, with its sculptured scales; *Pterichthys* is in the right-hand top corner; *Cephalaspis* in the left-hand bottom corner (with a distant one just above *Holoptychius*); in the left-hand top corner is *Pteraspis*, with *Glyptolepis* just below it, and the head of *Coccosteus* peeping out further down. *Osteolepis* is seen in the middle of the picture. It is not known whether all these different forms lived together at exactly the same time, but at least they all belong to the Old Red Sandstone, and for all we know may have been contemporaries.

At the present day both ganoid and plagiostomous fishes are in the minority, especially the former order, while the teleosteans

[1] *The Old Red Sandstone*, p. 100.

FISHES OF THE OLD RED SANDSTONE PERIOD.

Pteraspis.
Glyptolæmus.
Coccosteus (Head of).
Cephalaspis.

Osteolepis.

Holoptychius.

Pterichthys.

PLATE II.

abound in our seas, as they did during the Tertiary era. Although the palæozoic types are by no means low down in the scale, yet it is clear that the "Law of Progress" holds good for fishes too, and that, as time went on, a higher order largely replaced the lower ones. In Cretaceous times fishes abounded.

Mount Lebanon is a famous locality for fossil fishes of the latter period, both Selachians (of the order Elasmobranchii) and bony fishes (of the order Teleostei). The Rev. Professor Lewis had for years collected most valuable specimens from Sahel Alma and Hâkel. His specimens are far more perfect than those figured by Pictet and other authorities. The first mention of fossil fish in the Lebanon is to be found in Joinville's *Histoire de St. Louis.* During the sojourn of the king at Sidon, in the year 1253, just before his return home from the Crusades, a stone was brought him, says Joinville, "which was the most marvellous in the world, for when a layer of it was lifted, there was found between the two pieces the form of a fish. The fish was of stone, but lacked nothing in form, eyes, bone, colour, or anything necessary to a living fish. The king demanded a stone, and found a tench within." Professor Lewis says of Hazhula, one of the Mount Lebanon localities, "I was able to be in the place only about two hours, and was sufficiently annoyed in that time by the clamour for 'backsheesh,' and by the demand, made by twenty at a time, to buy whatever was offered. But in the short time I was there, and notwithstanding the unfavourable circumstances in which I was placed while there, I obtained about fifty good specimens, picked up from road and hillside. The exact place from which the stone came was very near by, but I did not see and examine it, so excited and clamorous did the ignorant fanatical people become, threatening to break all I had collected, frightening my muleteer, etc., led on by a few who pretended that I was seeking treasures. It required considerable tact, and a great waste of Arabic, to get off with my donkey-load of specimens, and, as it was, I had to send the muleteer ahead, and bring up the rear myself with a faithful attendant."[1]

At the end of the fossil-fish gallery at South Kensington is a plaster cast (in a frame over the door) of a large and very perfectly preserved ganoid fish, from the lithographic stone of Solenhofen, in Bavaria, viz. the *Lepidotus maximus* (see Fig. 12). The artist has drawn this fish in Plate VII., representing it

FIG. 12.—A ganoid fish, *Lepidotus maximus*. From the Solenhofen Limestone. 1. A scale. 2. Teeth.

in company with two interesting extinct gavials that lived in Jurassic times.

The Sturgeon (*Acipenser*) is another example of a living ganoid fish, often seen in fishmongers' shops. Considering its long pedigree, we ought to look upon this fish with great veneration! To the palæontologist, such survivals are not only very interesting, but most helpful in throwing light upon the organization of those ancient types whose remains are but imperfectly preserved in the record of the rocks.

[1] *Geol. Mag.*, Dec. 2, vol. v. p. 217.

CHAPTER III.

ANCIENT SALAMANDERS.

"Slowly moves the march of ages,
Slowly grows the forest-king,
Slowly to perfection cometh
Every great and glorious thing."

ANON.

In our last chapter we spoke of some of the fishes of bygone days. Now the next great step in the upward progress of the animal kingdom was the transition from the fish to the reptile. These two classes are at the present time well marked and clearly separated from each other; but an idea of how the change took place, in the course of Evolution, may be gathered from a study of the important group, or subclass, which comes in between the two, viz. the amphibians, or batrachia of some naturalists. This important and interesting group form, as it were, a series of "missing links" between fishes and reptiles. They derive the latter name from the Greek word *batrachos*, signifying a frog. But we will speak of them as amphibia,[1] because they live both in water and on land; and during the early part of their lives swim in water like fishes, and moreover breathe as fishes do by means of gills. Afterwards they put away such childish things as gills, and, developing lungs, become air-breathers, and behave like ordinary reptiles.

Modern naturalists who accept the theory of Evolution, as most

[1] Greek—*amphi*, both; and *bios*, life.

do, regard the metamorphosis of the tadpole into a frog as a kind of object-lesson on the part of Dame Nature, given with a view to show us, in a general way and in a short space of time, how the wonderful change from the fish to the reptile took place.

We have seen that, by the time the great coal-forests of the Carboniferous period were flourishing, the fish class had attained a high state of development, and were to some extent foreshadowing the amphibian type of creation. Now, we know that when the New Red Sandstone was being laid down, the next great class, viz. the reptiles, had not only made a start, but had struck out in several lines, and were specialised in certain directions, as shown by their bird-like footprints on the Triassic sandstones. Would it not therefore be expected that we should find, in rocks of greater age than the New Red Sandstone, the remains of animals in some respects intermediate between fishes and reptiles? This expectation is fully realised by the presence in such strata (*i.e.* the coal-measures and Permian rocks) of the remains of quite a large number of amphibia. They also crop up again in the Trias, but after that period are not much in evidence until we come to Tertiary times.

The amphibian class, at the present day, is represented by the frogs, toads, and newts of our own country, and by the salamanders of Japan and the United States. But these are only a meagre remnant of a large number of families that lived ages ago.

Let us endeavour now to take a general view of the amphibian class as it was in the days of long ago, before the creation of man,—to look at their family tree and see what their ancestors were like. When this is done we shall perhaps have a little more respect for the frogs and newts that inhabit our ponds; for they come of a very ancient stock, and one which played an important part in primæval times, branching out here and there into strange and varied forms, as if they could not help themselves. In order to do this we must go a long way back into the world's history; and even then we shall still be some way off the roots of the family tree; or, in other words, we shall be in the dark about the first forms of amphibian life which appeared on the earth's surface.

To find the oldest known amphibian we must go back as far as those "dark ages" when a large portion of the surface of what is now Europe was overgrown with vast areas of dismal coal-forests, something like the mangrove swamps of which we read in books of African travel; they must have been dark compared to an English meadow on a summer day, and therefore the expression "dark ages" is all the more appropriate. Our knowledge of the earliest air-breathing creatures is no doubt small, but of late years it has been considerably extended, thanks to the labours of several English, American, and Continental workers, such as Owen, Huxley, Miall, Seeley, Gaudry, Fritsch, Cope, and others, who have all carefully studied and described the fossil remains submitted to them, to say nothing of the numerous eager collectors who have brought away from the bowels of the earth numerous specimens of great value. Their services, although of a more humble order, are quite indispensable; and it should be borne in mind that it is in the power of all of us, however ignorant we may be of comparative anatomy, and therefore unable to write learned papers for the Geological Society, to render valuable service to the cause of Science simply by using our eyes and not neglecting opportunities of collecting specimens wherever we may go.

But before proceeding it will be necessary to introduce a new name for our old amphibians. It is rather a long one, we admit, but useful, because it partly serves to describe them. Scientific men are said to be very fond of long Latin and Greek names, but such can hardly be avoided; and there is this to be said for them—that the names they use are much more exact than the popular ones by which plants and animals are commonly known, and, moreover, they often convey a great deal of information in a brief manner. Now, the name given to the creatures we are about to describe is *Labyrinthodonts*.

It was Sir Richard Owen who first accurately described the fossil amphibians, and they received this name from him. Let us see what the name means and how it applies. Sir Richard Owen's acquaintance with these remarkable forms of ancient life began in the year 1840, when he first examined certain teeth from the New Red Sandstone of Coton End quarry, Warwickshire. In external character these fragmentary teeth corresponded with those which had been previously discovered in Germany (at Würtemberg) by Professor Jager, and which had been called by him *Mastodon-saurus*. This name was not a happy one, because it suggested an association with an extinct elephant, the *Mastodon*,[1] with which this amphibian, it is hardly necessary to say, had no connection, real or imaginary. Owen examined the teeth of the fossils from Germany, as well as those from Warwickshire, and found that, when cut across into transverse sections for the microscope, they revealed a very remarkable and complicated structure; the whole of the internal portion was seen to be made up of a complex series of foldings, forming a peculiar structure, suggesting a labyrinth—the external layer of cement belonging to the tooth converging in numerous folds towards the pulp-cavity. And so the name *Labyrinthodonts* reminds us that all, or nearly all, the fossil animals included under the above general term possess teeth having this peculiarity. They are the Labyrinth-toothed amphibians (see Fig. 13).

At first sight it might perhaps seem to the uninitiated as if the internal and minute structure of a tooth were a matter of comparative unimportance; but it is recognised by all anatomists that teeth are highly important as indicating the place of an animal in the scale of being. In the present case, for instance, the possession of teeth with the labyrinthine structure indicates an affinity with certain ganoid fishes, which also possess

[1] *Extinct Monsters*, p. 217, new edit.

similarly infolded teeth. Again, there is a peculiarly low type of reptiles, known as the *Ichthyosaurus*, or "fish-lizard,"[1] of which the teeth show an approach to the same kind of structure. Some of the Labyrinthodonts retained the gills of their youth through life, instead of changing them for lungs, as frogs do.

But, if these facts tend to connect the Labyrinthodonts with fishes, there are others which tend to connect them with the reptiles above. Sir Richard Owen, by a bold piece of reasoning, arrived at the conclusion that certain huge and somewhat frog-like skulls,

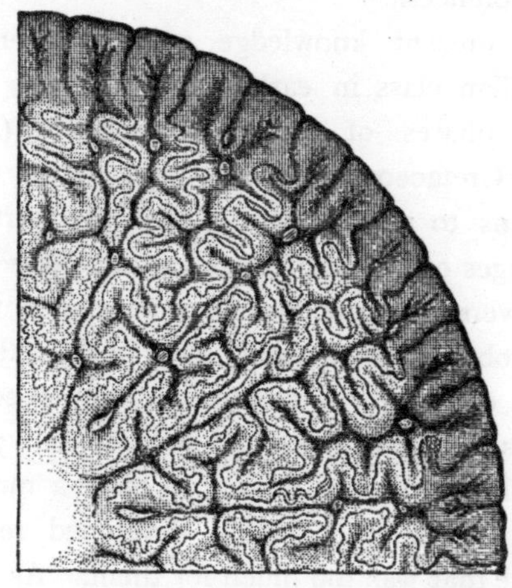

FIG. 13.—Section of tooth of *Labyrinthodon*. (After Owen.)

previously discovered in German Triassic rocks, belonged to Labyrinthodonts. This was an important step; but then the question arose, What kind of creature made the tracks found in these strata, both in Germany and England? This, as we have shown in chap. i., was decided in favour of the *Labyrinthodon* (previously known as the *Cheirotherium*, or "hand-beast").

Unlike the reptiles of to-day, modern amphibians have no fixed type of external form. Some, like frogs, suggest a likeness

[1] *Extinct Monsters*, p. 35, new edit.

to turtles and tortoises; others suggest in shape the true snakes; another group, represented by newts and salamanders, suggests a resemblance to the lizards; and lastly, those which keep their gills through life suggest certain finless fishes, such as the conger-eels and the lampreys. It was just the same in the old days, when amphibians flourished more abundantly than now; they struck out in many directions—some taking on a long snakelike form (see Plate III. and Fig. 22), others being short and with larger limbs, while in the matter of size there were great differences.

As far as present knowledge goes, it seems to indicate that the reptilian class in early days was less advanced than in the later phases of the Secondary era (*i.e.* during the Jurassic and Cretaceous periods); and so we can understand that amphibians to some extent played the part of reptiles in those remote ages of the coal-forests and the New Red Sandstone. Their very diversity shows that they were in a flourishing condition, and probably had a pretty wide field to themselves, without too many enemies. A certain amount of competition with other creatures, no doubt, they had to submit to; but with fishes they could hold their own. Later on came a much more severe form of competition from highly developed reptiles, such as Dinosaurs, and that was too much for them. As the reptile class rose in the scale of creation, amphibians, of course, came off second best, and the geological record tells us that, ever since, they have remained in the background of the theatre of life.

It will be well, before proceeding to describe some of the better-known types of Labyrinthodonts, to say a word about the general characters by which they may be distinguished. Of the internal structure of the teeth we have already spoken; but although the same complexity of folding does not apply to some of the smaller forms, the name by which they are known is very appropriate to the group as a whole. Another feature, revealed by their footmarks, is the marked difference in size between the hind limbs and the fore limbs, the latter being much smaller, as may be seen both in the figures of their skeletons and in those representing their tracks (see Figs. 2 and 23). There are five digits to each limb, and no claws.

Turning to their skulls, there is one feature that is quite peculiar to the group and unknown in any reptile, living or extinct, viz. that the whole of the upper surface of the skull

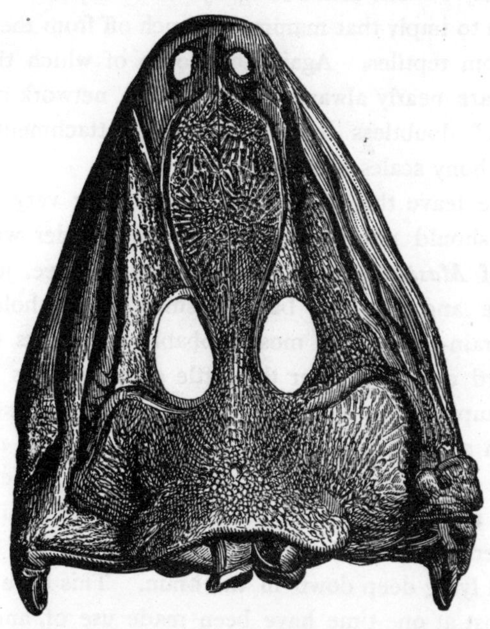

FIG. 14.—Skull of a Labyrinthodont, *Mastodonsaurus Jægeri*, from New Red Sandstone, Germany.

behind the eye-holes is covered in by a complete roof of bone. Hence some authorities call them *Stegocephali*,[1] or "roof-skulled." This is clearly shown in Fig. 14, representing the huge skull of *Mastodonsaurus*. Some fishes have skulls thus roofed over. But the skull of a lizard or a crocodile is very different, showing a long open channel behind each orbit, and

[1] Greek—*stegos*, roof; *cephalos*, head.

another one lower down at the side of the head. Another important character in the labyrinthodont skull is the presence of two condyles, instead of only one, as in all true reptiles and birds. The word *condyle*[1] signifies a knuckle, and is applied chiefly to the surface or surfaces by which the skull joins on to, or articulates with, the first vertebra of the neck. All the mammalia have skulls articulating by two condyles—a fact which would seem to imply that mammals branch off from the amphibia, and not from reptiles. Again, the bones of which the skull is composed are nearly always covered with a network of grooves, or "canals," doubtless intended for the attachment of hard, horny, and bony scales or scutes.

Before we leave the skull there is one other very interesting point that should not be omitted. If the reader will refer to the skull of *Mastodonsaurus* (Fig. 14), he will see, just on the middle line and near the base, a small round hole. It lies over the brain-cavity, and most probably represents the site of a small third eye. Whether this little extra eye was capable of receiving impressions of light, and so of communicating them to the brain as an ordinary eye does, is a matter of doubt; but in a certain very primitive little lizard now living in New Zealand, and known as the Tuatara (*Sphenodon*), there is a similar, though smaller aperture, overlying the rudiment of an eye now quite useless and lying deep down in the brain. This little eye in the Tuatara must at one time have been made use of, and the question arises whether the third eye of the Labyrinthodonts was similarly useless; but the chances are rather against the notion. All the lizards have this pineal eye. It is probable that we have here a relic from some much older and now lost form of life—possibly a fish, but possibly also a still lower type. These rudimentary structures, of which Darwin speaks in his great work, *The Origin of Species*, always possess a high degree of interest for the palæontologist, as being survivals from very early days in the

[1] Greek—*condulos*.

world's history. To the latter they are as interesting in their way as are ancient customs and old superstitions to a student of folk-lore or of ancient history.

We must also notice the forward position of the creature's nostrils as indicated by the two small openings at the apex of the skull (see Fig. 14). If we were to examine the under side of a skull such as that of the *Mastodonsaurus*, we should see that the bony palate is formed chiefly of two broad and flat bones, called the "vomerine," which generally support teeth. Now, this is a character not exhibited by the skull of any true reptile, and we must go to modern amphibia, such as our frogs and toads, to find a parallel. There is also a corresponding series of small teeth on the mandible of the lower jaw (see Fig. 24). The eyes often show the peculiar sclerotic plates seen so well in the "fish-lizards."

In studying an extinct vertebrate, or backboned animal, one of the first points to be considered is the nature of the vertebræ, or joints of the backbone. Now, in all fishes, and also in their distant cousin the *Ichthyosaurus* (fish-lizard), these bones have their articulating surfaces (centra) hollow, hence the popular term "cup bones" applied to the vertebræ of the latter. Need we be surprised, then, to learn that the Labyrinthodonts possessed similarly cup-shaped vertebræ? This is one more link to connect them with fishes. In some of the more primitive and ancient forms, the body of the vertebra was composed of three separate pieces, an arrangement peculiar to certain primitive fishes.

Ever since an unfortunate restoration of a Labyrinthodon by Mr. W. Hawkins, it has been the custom—in popular works at least—to represent the creature with a frog-like aspect. This restoration was based on the imperfect material of a good many years ago, and is therefore out of date. In our illustration, representing some of Mr. Waterhouse Hawkins's models in the grounds of the Crystal Palace (Fig. 52), the reader will see, far away to the right, a squat and toad-like creature; this will serve to give an idea of what that now historic restoration was like. But later discoveries have shown that, instead of being frog-like, most of the Labyrinthodonts, if we could have seen them in the flesh, would have reminded us more of our modern little newts, or of the salamanders of hot countries. The tail was generally well developed, and the limbs were adapted quite as much for walking as for swimming.

Their chests were defended by three bony plates (see Fig. 15), which had grooves similar to those on the skull. But besides this, all the under side of the body was protected by a large number of scales, running in two directions, as shown in Figs. 15 and 22. In one or two instances the whole body was covered with a coat of mail, consisting of small bony scales. An example of this kind of skin is the *Seeleya*, from the Permian rocks of Bohemia, restored by Fritsch, and named by him after Professor H. G. Seeley, F.R.S. In size the Labyrinthodonts varied greatly, some being only an inch or two in length, others seven or eight feet, and perhaps more. Most of them had limbs, but in one or two cases these had apparently been dispensed with.

It might naturally be asked, Did they undergo a series of changes similar to the metamorphoses of a common frog? One could hardly, perhaps, have expected an answer to this question; but, strange to say, some of the smaller forms from the petroleum shales (Gaskohle) of Bohemia, so carefully studied and described by Dr. Fritsch, have been very well preserved in those deposits of the Carboniferous age, so that he was able to recognise in some cases the signs of the gills, by means of which, in early life, they breathed in water, like fishes. And so it is manifest that they did go through a series of metamorphoses.

One of the most primitive, and at the same time one of the earliest known Labyrinthodonts, is the *Archægosaurus*, of which we must now give a short account.

The remains of this lowly amphibian—the humblest of all the Labyrinthodonts—were first discovered in certain clay-ironstones of the Lower Permian age, and also in the coal-field of Saärbruch, near Treves. At first they were supposed to represent fishes, but another specimen was recognised by Dr. Gergens as belonging

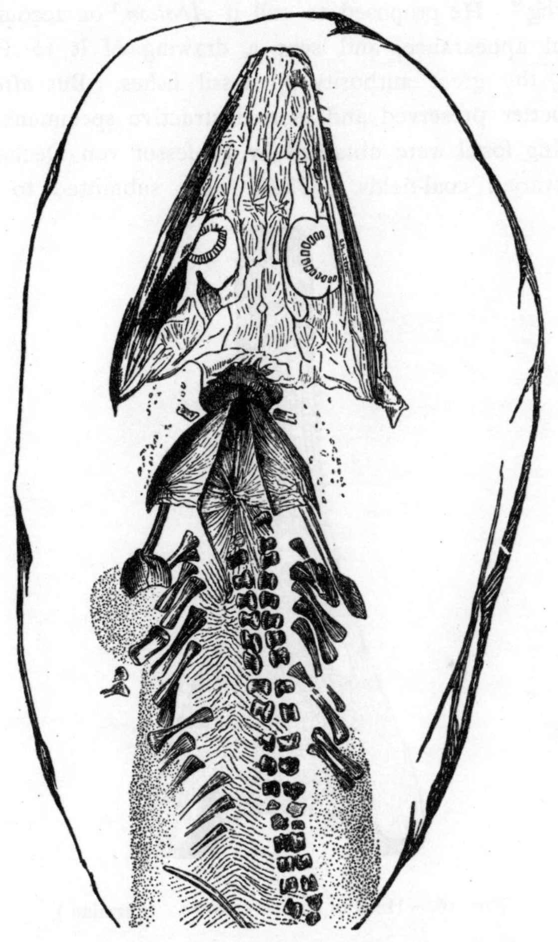

FIG. 15.—Part of skeleton of a Labyrinthodon, *Archægosaurus* (Decheni), from the Permian, Säarbruch. (After H. von Meyer.) The snout is incomplete. The three ventral plates are seen near the head.

to a salamander-like amphibian. This was sent to the celebrated H. von Meyer for description, who said, "Its head might be that of a fish as well as that of a lizard, or of a batrachian frog." He proposed to call it *Apateon*,[1] on account of its deceitful appearance, and sent a drawing of it to Professor Agassiz, the great authority on fossil fishes. But after three years, better preserved and more instructive specimens of this perplexing fossil were obtained by Professor von Dechen, from the Bavarian coal-fields. These were submitted to another

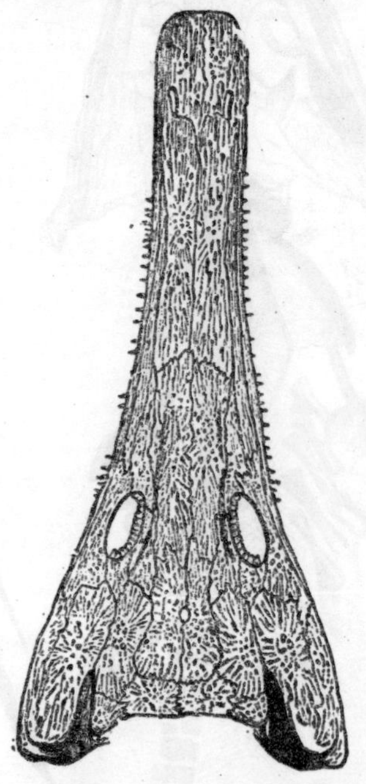

FIG. 16.—Head of *Archægosaurus*. (Permian.)

authority, Professor Goldfuss, of Bonn, who published a memoir thereon, with good figures. His opinion was that the remains belonged to a saurian genus, which he called *Archægosaurus*,[2] or

[1] Greek—*apateon*, cheat.
[2] Greek—*archégos*, leader; *sauros*, lizard.

"primæval lizard," deeming it to be a kind of "missing link" between the fish-like batrachia and the lizards and crocodiles.

Excellent casts of the specimens, figured and described, were sent by Goldfuss to Sir R. Owen, who placed them in the Museum of the Royal College of Surgeons. One of these appeared to the latter to show evidence of having possessed permanent gills, while the vertebræ and numerous very short ribs, with indications of stunted limbs adapted for swimming, all impressed him with the conviction that *Archægosaurus* was nearly allied to the modern *Proteus* and other amphibians that retain their lungs throughout life.

Some of the specimens acquired since Sir R. Owen published these conclusions are now in the Natural History Museum. There are certain fishes of the present day that have their backbones in an incomplete or unfinished state; that is to say, the bones consist partly of cartilage, like an unfinished house which is only built up to a certain level, the upper stories being merely indicated by scaffolding. An example of this kind of fish is furnished by the existing mud-eel, or *Lepidosiren*, a highly interesting form of life to the geologist, because it tends to fill up the gap between the two distinct classes of Fishes and Reptiles. Now, in this respect, the old Carboniferous amphibian we are now considering resembles the mud-fish. The latter leads a partly amphibious life, burying itself in the mud during the hot season, and breathing air by means of its air-sack. It is believed by evolutionists that, in the course of ages, lungs were developed from the air-sacks of "ganoid" fishes.[1] After detailing certain other characters, in which *Archægosaurus* resembles the higher ganoid fishes, Sir Richard Owen says, "All these characters point to one great natural group, peculiar for the extensive gradations of development, linking and blending together fishes and reptiles within the limits of such group." The salamander-like ganoids, *Lepidosteus* of North America (see p. 30) and *Polypterus*

[1] See p. 45.

of the Nile, are the most fish-like; the true Labyrinthodonts are the most reptilian or saurian of the group; *Archægosaurus* conducts the march of development from the fish proper to the labyrinthodon type; *Lepidosiren* conducts it to the batrachian type with permanent gills. Both tend to show the artificial nature of the arrangement by which, under most systems of classification, fishes are separated by a wide gap from reptiles.

In describing the amphibians of past ages it will be advisable to keep to our usual plan of taking them in chronological order, according to the accepted divisions of geological time, so that we may be able to compare those of one period with their descendants of another.[1] The Labyrinthodonts ranged through three geological periods—the Carboniferous, the Permian, and the Triassic. At the time when the coal-forests of Europe and America were flourishing, they were tenanted by a large number of Labyrinthodonts. These, and others of the succeeding Permian period, which may conveniently be taken together, have been specially studied by such authorities as Professors Huxley and Cope, Dr. Fritsch, Professor Miall, and others. The limits of space will only allow us to make a small selection from the numerous forms described by them.

In the year 1852 Sir Charles Lyell and Mr. (now Sir William) Dawson, who were investigating the coal strata of Nova Scotia, discovered an interesting little Labyrinthodont known as *Dendrerpeton.*[2] Its remains were found in the hollow of the trunk of a big fossil tree, the *Sigillaria*, which was completely converted into coal. The bones were considered by Professor Wyman, of Boston, to represent an amphibian, and on his advice they were brought to England and submitted to Sir R. Owen, who gave the above name to the creature to commemorate its discovery in a tree.[2] Another genus from the coal-measures of North America is *Raniceps* (frog-headed),[3] named by Professor Wyman. Its skull, fore

[1] See Appendix I. [2] Greek—*dendron*, tree; *erpeton*, reptile.
[3] Latin—*rana*, frog; *caput*, head.

limbs, and part of the backbone were found in a seam of coal in the great coal-field of Ohio, at the mouth of the Yellow Creek.

Some interesting footprints have been discovered in American coal-measures. In the year 1844, Dr. King, of Greensburg, Pennsylvania, discovered some tracks which he considered to be those of a reptile, in coal strata near that town. The impressions were stated by him to be nearly eight hundred feet beneath the topmost stratum of the coal formation. Sir Charles Lyell, on visiting Greensburg, examined the footmarks and confirmed Dr. King's view of them. He considered that they resembled the *Cheirotherium* footprints described in chap. i. "They consist," he says, "of the tracks of a large reptilian quadruped[1] in the sandstone in the middle of the Carboniferous series—a fact full of novelty and interest; for here in Pennsylvania, for the first time, we meet with evidence of the existence of air-breathing quadrupeds, capable of roaming in those forests where the *Sigillaria*, *Lepidodendron*, *Caulopteris*, *Calamites*, ferns, and other plants flourished." No less than twenty-three footsteps were observed, and those made by the hind limbs were larger; but these tracks only showed four toes, instead of the five of *Cheirotherium*.

Very similar footprints were discovered and described by Mr. Isaac Lea, in some red shales at the base of the coal-measures at Pottsville, north-east of Philadelphia. They are of older date, being seventeen hundred feet lower down. The name *Sauropus* has been given to the unknown animal that made them. Sun-cracks and the impressions of rain-drops were associated with them, showing clearly that the creature was an air-breather.

Cricotus is an American genus described by Professor Cope, from the Permian strata of Texas. The head and some of the scales belonging to the under surface of the body are shown in Fig. 17. The *Eryops*, of which the skull is shown in the same figure, is considered by Professor Seeley to belong to a higher,

[1] Such an expression as "reptilian quadruped" is not a happy one, and the term *quadruped*, as usually employed, may be replaced by the word *mammal*.

though related group, the *Anomodonts* (chap. iv.). The one group is much connected with the other, and the question of the

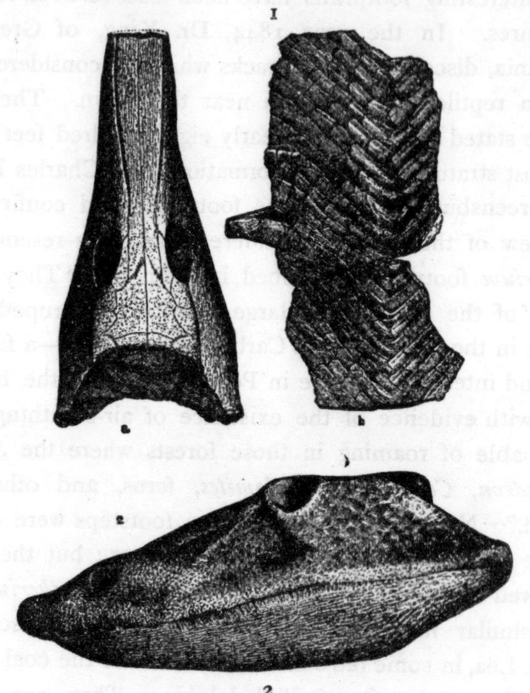

FIG. 17.—Labyrinthodonts. 1, *a*, *b*, *Cricotus*. 2. Head of *Eryops*. From Permian strata, Texas. (After Cope.)

classification of their various members is a difficult one, which only time and further discoveries can settle.

Loxomma is another genus of the Coal period, but only the head is known (see Plate III.). It was named by Professor Huxley, who, during a visit to Edinburgh in 1862, made a study of the large collection of vertebrate fossils from Burdie House and Gilmerton in the Museum of Edinburgh University. While looking through the collection for fossil fishes, he came upon some specimens which were of quite a different character, and

AMPHIBIANS OF THE COAL-FOREST PERIOD.

Actinodon. *Ceraterpeton.* *Archægosaurus.* *Dolichosoma.* *Loxomma* (Head of).

PLATE III.

among these found part of the skull of a Labyrinthodont, which he named *Loxomma*. A very beautifully preserved specimen of

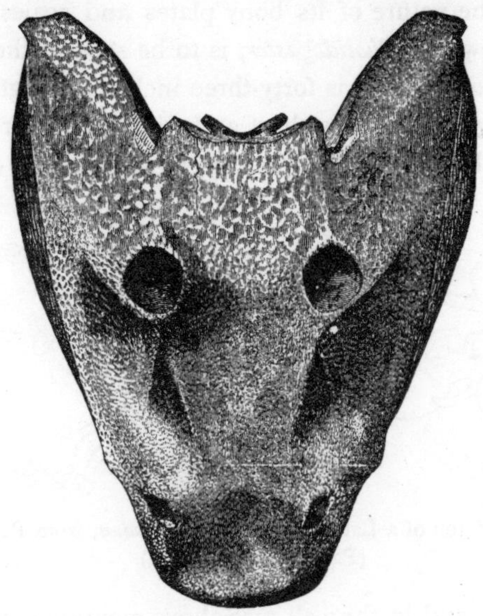

FIG. 18.—Head of *Eryops macrocephalus*, from Permian strata, Texas. (After Cope.)

the skull of this creature, from the coal-measures of Shropshire, is to be seen at the Natural History Museum (Gallery No. 5 of the Geological Department, Wall-case 11). It is fortunately uncrushed, and shows the natural contour of the skull and lower jaw. The specimen was presented by George Maw, Esq., F.G.S.

Sir Philip Egerton and the Earl of Enniskillen obtained from the Edinburgh district a remarkable skeleton, which looked very much like that of a fish; but after mature consideration they handed it over to the British Museum. Later on, Professor Huxley's attention was drawn to this specimen by Mr. Davies, of the Museum, who justly remarked upon a certain resemblance in the arrangement of its scales to the *Archægosaurus*. A careful study of the fossil by Professor Huxley bore out the suspicions of

Mr. Davies, and convinced him of the amphibian nature of the fossil. It differed from *Archægosaurus* in the form of its head, as well as in the nature of its bony plates and scales. This fine specimen, known as *Pholidogaster*, is to be seen in the same wall-case with *Loxomma*, and is forty-three inches in length.

In *Actinodon* the skull is short and wide, the nostrils are large, and the muzzle is broad. Fig. 19 is a drawing of the com-

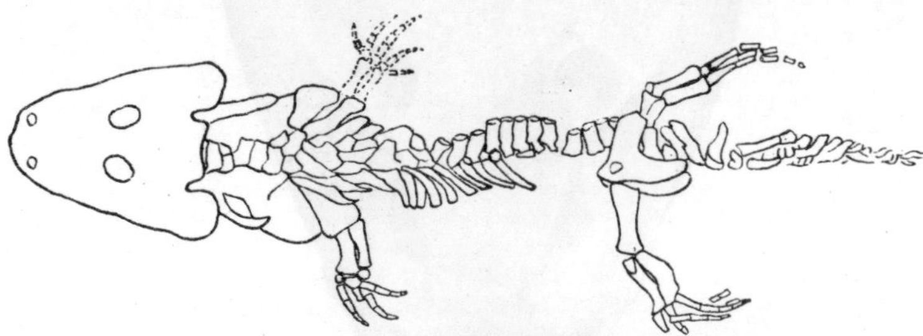

FIG. 19.—Skeleton of a Labyrinthodont, *Actinodon*, from Permian strata. (Partly after Gaudry.)

plete skeleton, and in the wall-case above mentioned will be found a skull preserved on a slab of shale from Permian strata in France. A restoration of the creature when alive is shown in Plate III.

A number of small Labyrinthodonts, somewhat like salamanders, have been found in Ireland and in Bohemia. Those from the latter country are very exhaustively described by Dr. Fritsch, in a valuable monograph,[1] with many excellent illustrations. We have reproduced two of these, the *Ceraterpeton* and the snake-like *Dolichosoma*[2]—both restored in Plate III., and their skeletons shown in Figs. 20, 21. Speaking of snake-like forms, it may be mentioned that one of them—the *Palæosiren* ("ancient siren")—is estimated to have attained a length of forty-five feet. Professor Cope has written an account of a number of Labyrinthodonts from the coal-measures of Ohio, where both Professor

[1] *Fauna der Gaskohle*, Dr. A. Fritsch.

[2] Greek—*dolichos*, long; *soma*, body.

Newberry and Professor Wyman have obtained specimens. These show great variety of form. Some resemble long-limbed lizards;

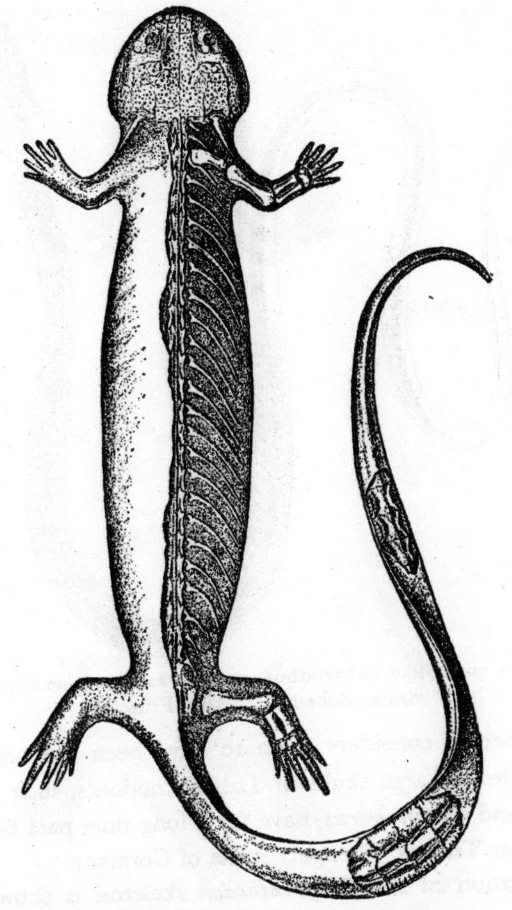

FIG. 20.—A small Labyrinthodont, *Ceraterpeton*, restored after Fritsch, from Permian strata, Bohemia.

some have short limbs. Others, like *Ptyonius* (see Fig. 22), were very elongate. The members of the Labyrinthodont order flourished over a large part of the earth's surface, from the time of the coal-forests to that of the Triassic sandstones. Their remains are found in all the great continents, except South America.

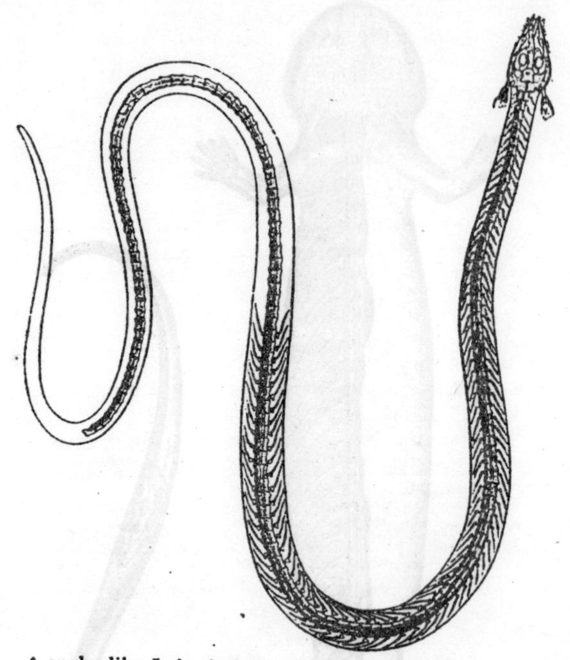

FIG. 21.—A snake-like Labyrinthodont, *Dolichosoma*, from Carboniferous strata, Bohemia. (After Fritsch.)

Professor Seeley considers them to have been the ancestors of the crocodiles. Large skulls of Labyrinthodonts, such as *Mastodonsaurus* and *Capitosaurus*, have for a long time past been found in the Upper Triassic (Keuper) strata of Germany.

A restoration of the *Labyrinthodon* skeleton is shown in Fig. 23, and we must now say a few words about the specimen on which this valuable restoration has been made.[1]

In the year 1877, M. Rütimeyer wrote to Professor Wiedersheim to say that the Museum at Basle possessed a specimen of

[1] A complete account is given by Professor Wiedersheim in *Abhandlungen der Schweizerischen Palaontologischen Gesellschaft*, vol. v. (1878).

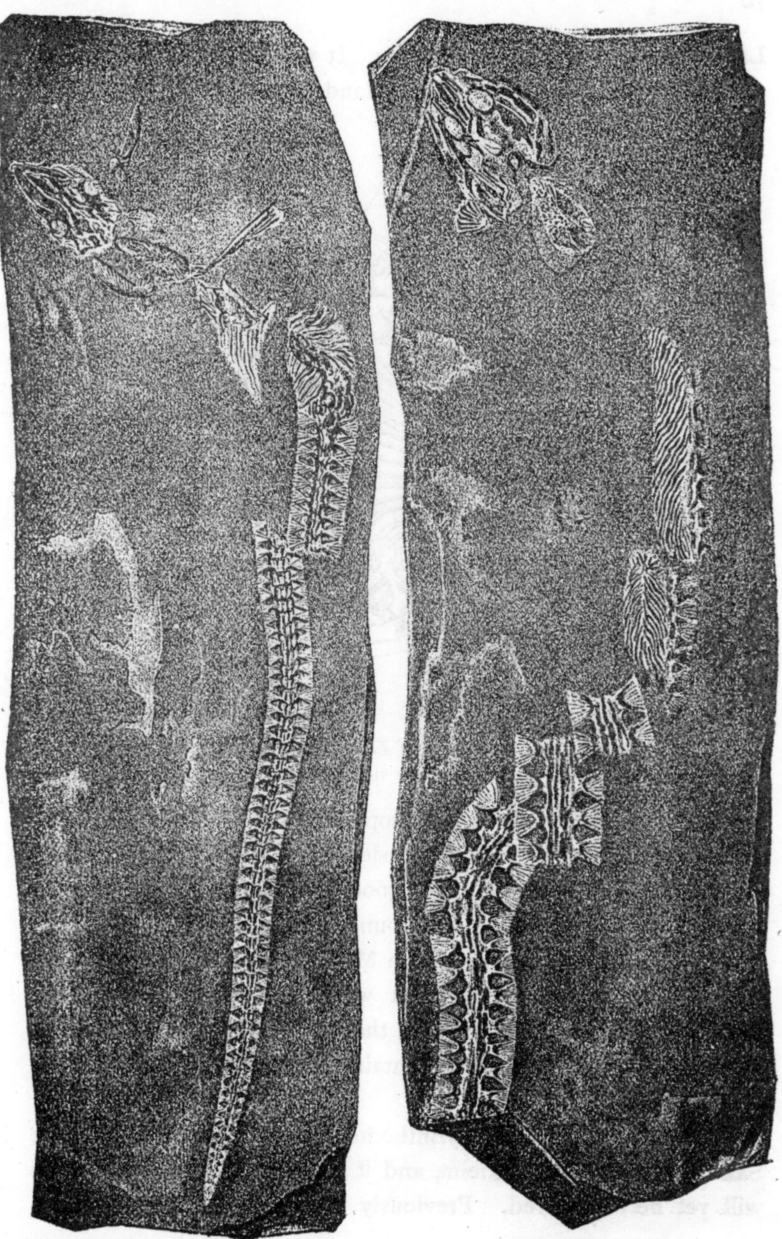

FIG. 22.—*Ptyonius* (two species). (After Cope.) From coal-measures, Ohio.

Labyrinthodon in good condition. It was found in 1864, in a sandstone quarry in Riehn, Switzerland (Riehn is the first station

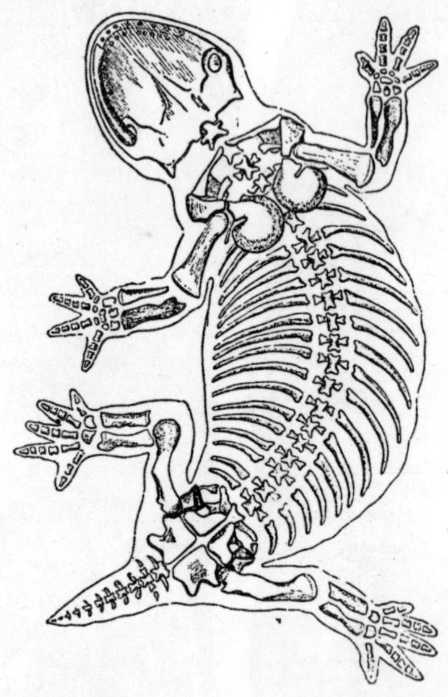

FIG. 23.—Restoration of skeleton of *Labyrinthodon* (after Wiedersheim), *under* side.

on the railway from Basle to Schopfeim), and was placed in the collection of the University of Basle. The finder was a M. Frey, an architect, who, in spite of a good offer from Stuttgart, refused to allow his fossil to leave the country. Later on, a photograph of the specimen was sent to Von Meyer, but for some reason he did not describe it. The task was therefore undertaken by Professor Wiedersheim. Riehn, the place of discovery, lies on a minor range of the Vosges Mountains bordering the Black Forest on the south.

Detached plates of Labyrinthodonts are found in the Bunter Sandstone quarries of Riehn, and it is likely that other skeletons will yet be discovered. Previously to this find, only fragments

REPTILES AND AN AMPHIBIAN OF THE NEW RED SANDSTONE PERIOD.

Rhynchosaurus. *Mastodonsaurus.* *Hyperodapedon.* *Telerpeton.*

PLATE IV.

of *Mastodonsaurus* and *Capitosaurus* were known. As is usual with skeletons in sandstone, the bones, having been dissolved away, are mostly represented by casts in the rock. When first found, the slab of rock containing this fossil only revealed a part of the skull of the creature, so that much labour and care were required to "develop" the specimen. The reader should be careful to note that Fig. 23 represents the *under* side of

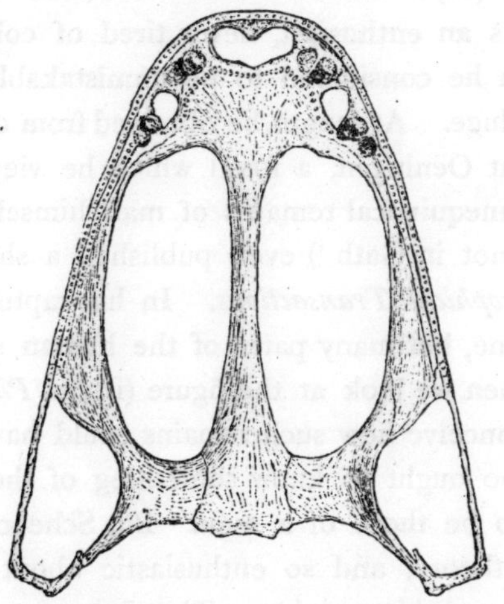

FIG. 24.—Lower jaws of a Labyrinthodont, *Cyclotosaurus*.

the skull, and so what looks, at first sight, like an eye, is only a feature of the under side of the skull. Professor Wiedersheim was able to make a study of the brain, and found it to be remarkable for occupying so large a space in the skull; but it is of a low type. It is curious to find in this Labyrinthodon an entire absence of bony plates and scales—even the usual three plates on the chest being absent. The tail, also, is unusually short; and it would seem that this creature was less aquatic in its habits than some of the other forms we have been considering. It was not a large animal, the length of the skeleton on the slab being under two feet.

Leaving now the Labyrinthodonts—which we have called "Primæval Salamanders," for want of a better name, as there is nothing like them alive now,—we pass on to say a word or two, in conclusion, on the remains of a true salamander, discovered in Tertiary strata on the Continent.

Few fossils have awakened more curiosity than the skeleton of Scheuchzer, who was rash enough to call his specimen *Homo Diluvii Testis* (or, "the man that witnessed the Deluge"). Scheuchzer was an enthusiast, never tired of collecting organic remains, which he considered to be unmistakable evidences of the general Deluge. At length he obtained from certain strata of Miocene age, at Oëningen, a fossil which he viewed with great delight as the unequivocal remains of man himself. The Royal Society (tell it not in Gath !) even published a short description in their *Philosophical Transactions.* In his rapture, Scheuchzer saw not only one, but many parts of the human skeleton in his specimen. When we look at the figure (in his *Physica Sacra*), it is difficult to conceive how such remains could have appeared to a physician, who ought to know something of the bones of the human body, to be those of a man. But Scheuchzer was so in love with his theory, and so enthusiastic about it, that many naturalists adopted his opinion. The first man who dared to doubt was Gesner (1758), who had a specimen of his own. A third specimen of this fossil came into the possession of Dr. Anman, of Zurich, and is now in the Natural History Museum (Gallery No. 5, Wall-case 11). Cuvier well observed that a comparison of the skeleton with that of man would at once have dispelled Scheuchzer's idea, which was consequently abandoned. The head and large orbits for the eyes struck Cuvier as strongly suggesting the head of a frog, or of a salamander. When staying at Haarlem in 1811, he obtained permission to work at the stone, and thus found more of the bones; in this way he settled the question, and proved that the supposed man who saw the Deluge was only a salamander after all!

CHAPTER IV.

ANOMALOUS REPTILES.

"In the endeavour to complete the Natural History of any class of animals, the mind seeks to penetrate the mystery of its origin, and, by tracing its mutations in time past, to comprehend more clearly its actual condition, and gain an insight into its probable destiny in time to come."—SIR R. OWEN.

HAVING, in our previous chapter, discussed the Labyrinthodonts, we now pass on to consider some of their descendants in the shape of a very peculiar group of reptiles hailing from South Africa and elsewhere. Perhaps the chief peculiarity, among many others, presented by these very antique and old-fashioned inhabitants of the world, is to be found in their teeth. In fact, so irregular and anomalous are they in regard to these organs, and so at variance with all our preconceived ideas with regard to what proper, well-behaved reptiles, whether living or extinct, ought to be like, that their vagaries in this respect have led to their being christened by the name of *Anomodonts*, or "anomalously toothed" reptiles.[1] Needless to say, they occupy a distinct order in the classifications adopted by palæontologists. The moral of all this is, as we shall better understand later on, that it would be well for students of extinct forms of life to enter this domain of Science without any preconceived ideas at all! It would save a great deal of confusion and trouble in the end; and, moreover, would be far more truly scientific. For what right has any one, however great his knowledge or his ability, to dictate to Nature,

[1] Greek—*anomos*. without law: *odous*. *odontos*. tooth.

and to say this or that is impossible—that no reptile, for instance, could possibly have flown; or that such and such teeth were impossible for a reptile?

We now know that there was a time when certain reptiles *did* fly (although many people with some pretensions to knowledge doubted the evidence). And so with regard to reptilian teeth; fossil evidence shows that some old reptiles had teeth more like those of modern mammals! Facts such as these should teach caution, and every student of palæontology will do well to remember the saying of Agassiz: "The possibilities of existence run so deeply into the extravagant that there is scarcely any conception too extraordinary for Nature to realise."

The chief characters of the Anomodonts may be briefly stated as follows: In this order the body is lizard-like, and the limbs are adapted for walking. The skull is comparatively short, and the nasal openings are large. The teeth are generally placed in distinct sockets (thecodont). The bodies of the vertebræ are hollow at both ends (amphicœlous), and in some cases are only partly converted into bone—a character which is common to the Labyrinthodonts. The whole structure of the foot is distinctly on the mammalian plan. Recent researches show that these animals are descended from Labyrinthodonts, and more especially the family of which *Archægosaurus* is a member (see p. 59). Certain important characters show (strange as it may seem) affinity with mammals; and it is probable that they are related to the lowest group of them, as represented at the present day by that remarkable creature the Spiny Ant-eater of Australia (*Echidna*), and the wonderful Duck-mole (*Ornithorhynchus*), which lays eggs like a reptile. These two creatures belong to the Monotreme order. We shall have more to say presently about the points of resemblance between Monotremes and Anomodonts (see p. 86).

Now the Anomodont reptiles are divided into several groups sub-orders, and families, of which we will take first those known as *Dicynodonts*,[1] because they illustrate the anomalous nature of the teeth, to which we referred just now.

The singular genus *Dicynodon* was discovered by Mr. Andrew G. Bain, in South Africa. In the year 1844 this enthusiastic collector, who had been engaged in the construction of military roads in the Colony of the Cape of Good Hope, discovered, in the tract of country extending northwards from the county of Albany, about four hundred and fifty miles east of Cape Town, several nodules, or lumps of a kind of sandstone, which, when broken, displayed, in most instances, evidences of fossil bones, and usually of a skull with two large projecting teeth. These fossils were first made known to English geologists under the name of "Bidentals," by Mr. Bain, on account of their two teeth, or tusks, and were sent to Sir R. Owen for examination and description. The specimens were exhumed by Mr. Bain from the intensely hard nodules of the sandstone strata which range over an immense tract of country beyond the mountains north of Cape Town, and may now be seen in the Fossil Reptile Gallery of the Natural History Museum.

Sir R. Owen, after a careful study of these interesting remains, concluded[2] that there had formerly existed in South Africa—probably in a great lake or inland sea—a race of reptiles presenting, in the construction of their skulls, characters presented by the crocodile, the tortoise, and the lizard, but possessing a pair of long tusks implanted in distinct sockets, like those of the walrus. No other kind of teeth had they; and, as in the case of a tortoise, the lower jaw appears to have been armed by a trenchant sheath of horny matter. The tusks are of a finer texture than that of the crocodile's teeth, and almost as dense as in the hyæna. We have here, then, a singular order of ancient reptiles, presenting

[1] The genus *Dicynodon* is so called from two Greek words: *dis*, twice; and *kunodos*, dog-toothed, on account of the two tusk-like canine teeth in the upper jaw.

[2] Professor Owen's *Memoir on the Dicynodon*, *Geol. Trans.*, second series. vol. vi., with plates.

in a most striking manner that blending of the peculiarities of several existing orders which is continually presented to the palæontologist. The head is shown in Fig. 25, and the fore limb in Figs. 26, 27. As it is, at present, hardly possible to restore the skeleton with any degree of certainty, we have not ventured to show the Dicynodon in our Plate of New Red Sandstone Reptiles (see Plate IV.).

Some of the arguments used by Owen in his *Memoir*, above

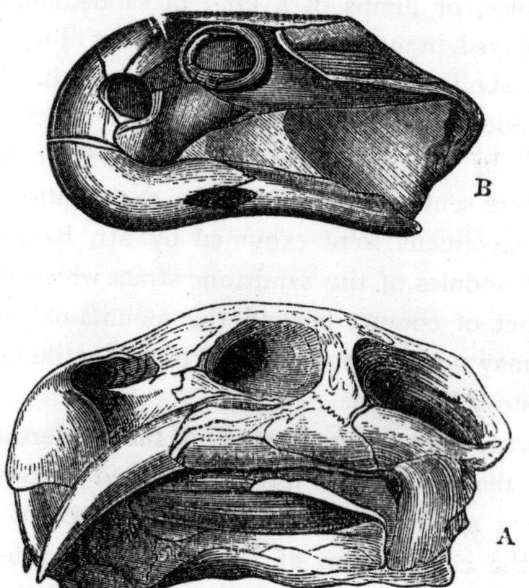

FIG. 25.—Anomodont skulls. A, head of *Dicynodon;* B, head of *Oudenodon*, from Karoo strata, South Africa. (After Owen.)

referred to, may be briefly condensed as follows: The creatures to which these skulls belonged were not mammals (although to an evolutionist they are a foreshadowing of that class), both on account of the double nasal apertures, one of which is seen in Fig. 25; and also because no mammal has a brain-cavity so relatively small. They were not crocodiles, as indicated by certain other features of the skull. They were not turtles or tortoises, for all such reptiles have a single nasal opening placed in the middle of the fore part of the skull. They could not be fishes, for fishes breathe in quite a different way. Neither could they be batrachians (frogs, etc.), nor yet snakes, as is also proved by the structure of their skulls. Certain other features of the skull show a relationship with the Lacertilians, or true lizards. The skulls are mostly of small size; but that of *Dicynodon tigriceps* is as much as twenty inches in length. That of *D. lacerticeps* ("lizard-headed"), Fig. 25, is only six inches long.

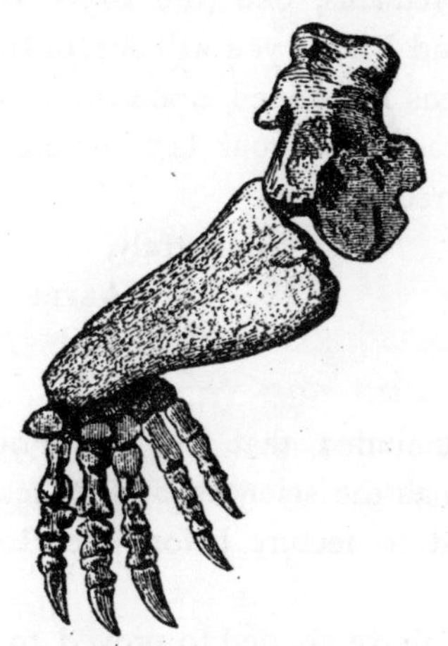

FIG. 26.—Fore limb of *Dicynodon*. (After Owen.)

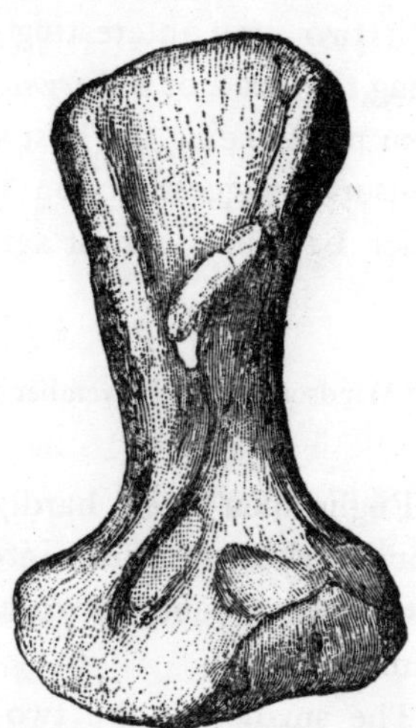

FIG. 27.—Arm-bone (humerus) of *Dicynodon*. ¼ natural size. (After Owen.)

Dicynodonts have since been found in the Gondwana series in Central India, and in the Elgin Sandstone, Scotland.

The vertebræ, or joints of the backbone, are hollow on both sides (biconcave)—a feature common to fishes and Labyrinthodonts. Probably they were good swimmers, and spent much of their time in the water; but it is quite clear that they were air-breathers, and so must have come up to the surface to breathe.

It is not every palæontologist who can claim the co-operation of Royalty in obtaining fossils—at least since the days of Alexander the Great, who collected all kinds of natural history specimens for his teacher, Aristotle, in the many lands he visited during his campaigns,—but in this respect Professor Owen was highly favoured, as the following note from H.R.H. Prince Alfred (Duke of Saxe-Coburg), written in 1860, will show :—

"DEAR PROFESSOR OWEN,

"In the course of my journey in South Africa I met with two very interesting fossil remains, one (the larger one) being the head of a *Dicynodon ;* and I hope you will accept them from me as being the best specimens I obtained, upon the Prince Consort's suggestion, on the occasion of your last lecture, of which I retain the most agreeable recollection.

"Yours truly,

"ALFRED.

"Windsor Castle, November 15."

Englishmen need hardly be reminded that the late Prince Consort showed a keen interest in all the sciences, and Professor Owen was frequently commanded to lecture before the Royal Family.

The smaller of the two fossils above alluded to proved to be one of the most perfect specimens of a Dicynodont skull known to the Professor. The genus *Oudenodon*,[1] although it has no teeth, so closely resembles the *Dicynodon* that it must be included in the same family. The skull is shown in Fig. 25. Professor Owen even suggested that the absence of teeth might merely denote a difference of sex ; but this view is not accepted by others. The general shape of the skull is very similar to that of *Dicynodon*, and the jaws ended in a kind of beak. There are several species, and some attained very large dimensions.

In *Oudenodon Bainii* the skull is six inches long. Mr. Bain, in

[1] Greek—*ouden*, nothing ; *odous*, *odontos*, tooth.

a letter announcing the discovery of the fossils in South Africa, mentions that other bones were found in association with the skulls, which give additional evidence of the saurian nature of the toothless reptiles. He writes,[1] "There were many skulls entirely without teeth, which we at first thought had belonged to chelonians or turtles; afterwards, finding that the animals had distinct narrow ribs, which chelonians have not, we put them down also for something new, and named them 'Oudenodons,' or toothless animals."

Platypodosaurus[2] is the name given to a considerable portion of the skeleton of a Dicynodont reptile from the Karoo strata of South Africa; but, unfortunately, the skull is unknown, and it may prove to be identical with *Oudenodon*. The structure of the pelvis, or region of the thighs, is remarkably mammalian. The name indicates that it had broad feet.

The genus *Endothiodon* represents a remarkable family of large reptiles from the same strata, distinguished by the presence of teeth on the palate. The skull resembles that of Oudenodon, but the muzzle is longer.

Two genera, *Placodus*[3] and *Cyamodus*, from Triassic strata in Germany, represent another order, the Placodonts, which are very remarkable, but unfortunately at present are only known by their skulls. These, however, are sufficiently peculiar to deserve notice (Table-case 18). The teeth represent a curious modification previously unknown in the reptile class, but of which the class of fishes affords numerous examples. In the latter genus the skull is as broad as it is long, and the jaws were very strong. This powerful action of the jaws relates to the form and size of the teeth, which are broad and flat, like paving stones, and evidently adapted to crack and bruise shells of molluscs and crustacea. Although now admitted to be a reptile, this remarkable genus was once con-

[1] The *Eastern Province Monthly Magazine*, Graham's Town, September, 1856.

[2] Greek—*plàtus*, broad; *pous*, *podos*, foot; and *sauros*, lizard.

[3] Greek—*plax*, *plakos*, any flat thing; *odous*, tooth.

sidered by Agassiz, Owen, and others to be a fish; but where fossil remains are so imperfect, mistakes may easily be made at first.

Tapinocephalus[1] and *Titanosuchus*[2] are the names given to two huge Anomodonts from South Africa, both belonging to the same family, the teeth of which indicate carnivorous habits. An imperfect skull, several entire limb-bones, and vertebræ are preserved in the national collection at South Kensington.

Galæsaurus,[3] of which the head is shown in Fig. 28, belongs to a remarkable group of Anomodonts, first described by Sir R. Owen, and called by him *Theriodonts*,[4] because the form and order of arrangement of their teeth bear a striking resemblance to those of carnivorous mammals; for the incisors are separated from the molars by well-developed canines, and the canines of the lower

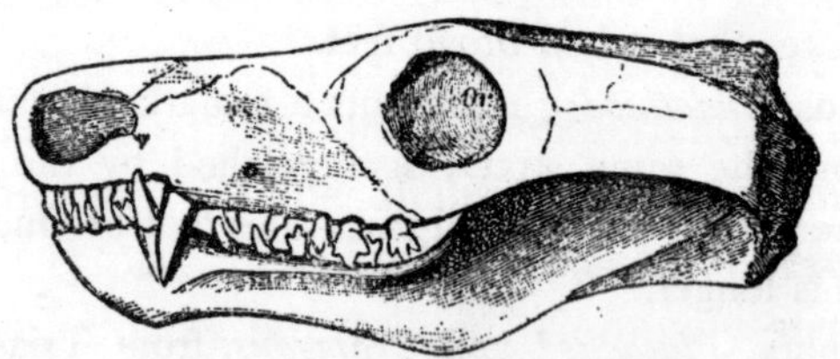

FIG. 28.—Head of an Anomodont, *Galæsaurus planiceps*. Length 6 inches. Karoo strata, South Africa. (After Owen.)

jaw crossed those of the upper in front. In some of the members of this family the upper canine teeth are long and trenchant, and the incisors large and close together. Visitors to the Natural History Museum will find a most valuable set of specimens of the skulls, etc., in Gallery No. 4, Table-case 19.

In most reptiles, living and extinct, the teeth that are worn away by use, or otherwise lost, are replaced by others that are constantly forming in the jaws; but in the case of *Theriodonts*

[1] Greek—*tapeinos*, low; *cephalos*, head.

[2] Greek—*titan*, a titan; *souchos*, a crocodile.

[3] Greek—*galé*, weasel; *sauros*, lizard.

[4] Greek—*therion*, wild beast; and *odous*, *odontos*, tooth: because the teeth resemble those of savage carnivorous creatures, such as lions or wolves.

there is no evidence of preceding teeth, like the milk-teeth in mammals, nor of succeeding teeth, like the crocodile's. Sir R. Owen therefore concluded that these creatures had but one set of teeth, which lasted through life. He has described eleven genera, varying in the size and form of the skull and teeth; they are all from South Africa, and are figured in his *Catalogue of the Fossil Reptiles of South Africa.*

The work of describing and classifying Anomodont reptiles has since been carried on by Professor H. G. Seeley, F.R.S., who has contributed a series of elaborate papers to the Royal Society.[1]

In America and France the subject has been studied by Professors Cope[2] and Gaudry. The former palæontologist has described, amongst others, such remarkable forms as *Eryops* (Fig. 18), *Empedias*, and *Dimetrodon*. As these are not completely known, we have refrained from attempting to restore them. Perhaps the most remarkable is the *Dimetrodon*[3] (see Fig. 29). Not only was the skull provided with formidable tusks, but the vertebræ of the back present a new feature in having very long neural spines. In one species the height of the spine is actually more than twenty times the length of the centrum! According to Professor Cope, these spines formed a kind of elevated fin on the back, of which it is difficult to imagine the use; but then there are many living animals with bony structures which, if only known in a fossil state, would greatly puzzle every one.

So far we have only spoken of Anomodonts that are imperfectly known; it therefore now remains, before we part company with this wonderful extinct order, to describe one of which the skeleton is practically complete. This is the remarkable *Pareiasaurus*,[4] for

[1] *Philosophical Transactions*, vols. 179, 183, etc.

[2] *History of the Vertebrata of the Permian Formation of Texas*, by Professor E. D. Cope, Pal. Bull, No. 32.

[3] Greek—*dis*, double; *metron*, measure; *odous*, tooth.

[4] Greek—*pareia*, the cheek-piece of a helmet; and *sauros*, lizard. So

the discovery and description of which palæontologists are indebted to a distinguished English geologist, Professor H. G. Seeley, F.R.S.[1] Visitors to the Natural History Museum will see the unique specimen brought home by Professor Seeley at the end of the fossil reptile gallery, in a glass case by itself (see Fig. 30). The

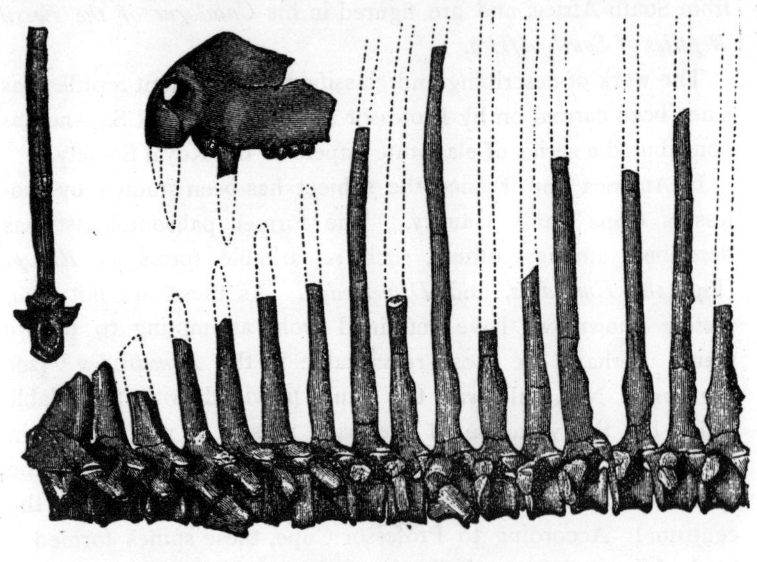

FIG. 29.—Parts of skeleton of *Dimetrodon incisivus*, from Permian strata, Texas. About ¼ natural size. (After Cope.)

story of its discovery, as given in the journals above quoted, may be briefly condensed as follows:—In the year 1889 Professor Seeley visited Cape Colony, and examined the Museums of Cape Town and Graham's Town, with a view to studying such remains

named because the cheek-bones descend so as partly to cover the back of the lower jaw.

[1] *Philosophical Transactions*, vol. 183 B (1892), and *Journal of South African Philosophical Society*, vol. vi. p. 5 (1889–90).

of the creature as were then known; but almost every specimen

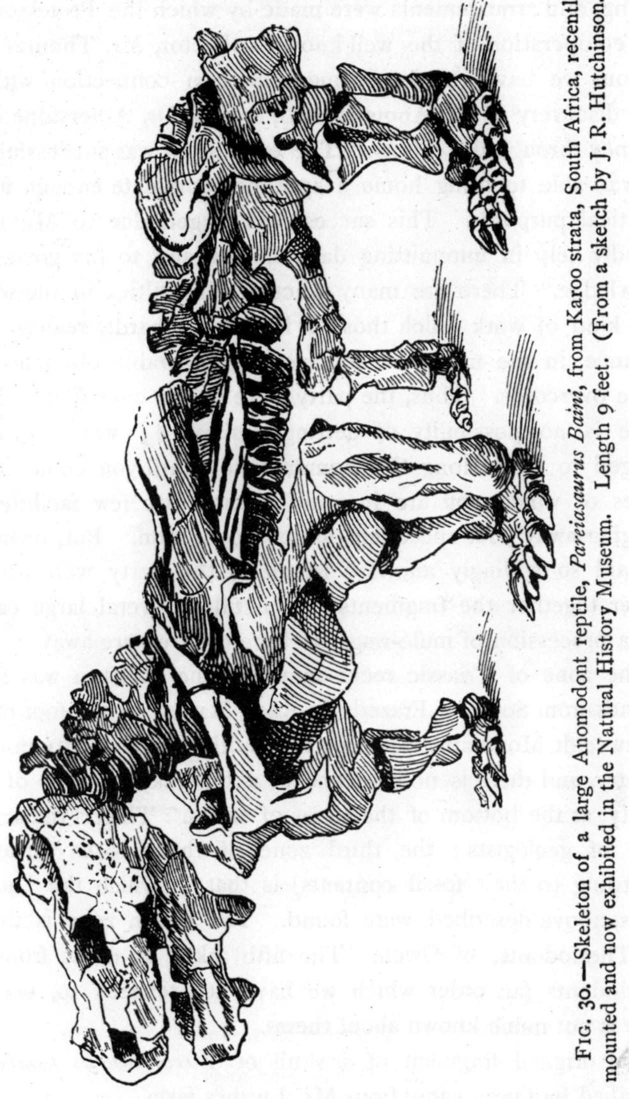

FIG. 30.—Skeleton of a large Anomodont reptile, *Pareiasaurus Baini*, from Karoo strata, South Africa, recently mounted and now exhibited in the Natural History Museum. Length 9 feet. (From a sketch by J. R. Hutchinson.)

of real interest had already been sent to the British Museum.

He therefore resolved to try and collect new material. Sir Gordon Sprigg kindly helped to insure the success of the undertaking, and arrangements were made by which the Professor had the co-operation of the well-known collector, Mr. Thomas Bain (whom we have previously mentioned in connection with the first discovery of the Anomodonts), and of Dr. Atherstone in his journey through the colony. The expedition was successful, and he was able to bring home a specimen complete enough for all practical purposes. This success was largely due to Mr. Bain's friendly help in unremitting daily labour, and to his great local knowledge. There are many practical difficulties in the way of this kind of work which those at home can hardly realise. For instance, in the matter of transport considerable obstacles have to be overcome. Thus, the party were in the open Veldt, where there is no possibility of getting assistance; where you are obliged to hunt along the mountain-side until you come to the bones of which you are in search; there are few facilities for bringing away your specimens in good condition. But, owing to the aid so willingly afforded to them, the party were able to gather together the fragments (which filled several large cases), and a procession of mule-waggons bore the treasure away.

The zone of Triassic rocks in which the skeleton was found extends from South of Frazerburg Road Station to the foot of the Nieuwveldt Mountains, covering a breadth of about fifty miles of country, and there is no evidence of a "break," or want of continuity, at the bottom of the series of strata. This is the second zone of geologists; the third zone of these rocks (arranged according to their fossil contents) is that in which the Dicynodonts, above described, were found. The fourth zone is that of the Theriodonts, of Owen. The fifth takes its name from the Zanclodonts (an order which we have not referred to, because there is not much known about them).

The original fragment of a skull of *Pareiasaurus bombidens* described by Owen came from Mr. Luttig's farm.

AN ANOMODONT REPTILE (PAREIASAURUS) FROM THE KAROO FORMATION, SOUTH AFRICA.

PLATE V.

The specimen (*P. bombidens*) described by Professor Seeley in the *Philosophical Transactions* for 1888, is from Palmiet Fontain, north-west of Tamboer, at the foot of the Nieuwveldt range. He accordingly selected this region of the Karoo system of rocks as a likely one for a find. Eventually, under the guidance of Mr. Serel Marais, he saw a large skeleton like that of a *Pareiasaurus*, on the side of a hill at a place called Bad. An expedition was at once organised, the result of which was that this fine specimen was quarried from the hillside on August 12, 1889, under the direction of Professor Seeley. Every block and every fragment was carefully marked with colour as it was removed, so that the pieces of stone might be eventually fitted together again. It became manifest during the work of excavation that the specimen included all the essential parts of the skeleton, and that it belonged to a massive animal with short, solid, and heavy limb-bones, with long ribs, and with a certain amount of armour in the form of bony plates or scales down the back. On arriving home, Professor Seeley fitted the fragments together at the Natural History Museum, and built up the whole specimen, matrix and all, into a solid mass. In January, 1890, Mr. Richard Hall, assistant mason in the geological department, began to remove the matrix, and has since developed the specimen with a skill, patience, and success which have never been surpassed. Few parts of the skeleton are absent, and nothing of any importance, except, perhaps, some of the ribs. The skeleton, as now mounted, is pretty much in the same position as that in which the bones were found in the rock, but the head is slightly raised.

There is probably no other South African fossil reptile in which the teeth are developed to the same extent in rows on the palate. Three rows of scutes, or bony scales, appear to have extended down the middle of the back. Some of these are still preserved in contact with the neural spines on the specimens, but they are not easily detected. Each scute is about two inches wide. There were also scutes on the skull, as in the Labyrinthodonts.

The creature must have been fully nine feet long when alive. We have with much diffidence ventured a restoration, as shown in Plate V. The great width and squareness of the animal is a feature that strikes one at once. The skull decidedly resembles that of a Labyrinthodon, and, even to a casual observer, has a frog-like look. But internally it is like that of the Tuatara, or *Sphenodon*, of New Zealand, which is a survival from very old times, and throws much light on several orders of Triassic reptiles. The backbone, or vertebral column, also resembles that of the little Tuatara, and not that of a Labyrinthodon. In the shoulder girdle we see a resemblance to the Anomodonts, as also in the pelvis, or region of the thighs, which is remarkably mammalian. Every one who looks at the skeleton must be struck with the great strength and massiveness of the limb-bones. The arm-bone suggests two very triangular wedges fastened together; so does the femur, or thigh-bone. For the sake of comparison, a specimen of the Tuatara is placed in a case close by, together with its skeleton (Wall-case 7). Visitors will find it instructive to compare this skeleton with that of the *Pareiasaurus*. On so doing, they will see a resemblance in the limb-bones of the two creatures. How fortunate that this little lizard should have survived so many geological ages—as it were on purpose to give us some help in studying the long-lost Anomodonts! Professor Seeley considers that the limbs, as represented by their bones, are intermediate between those of mammals and the tailed amphibians, such as newts and salamanders. But the creature was more of a reptile than an amphibian, as is shown by the fact that its skull articulated with the first joint of the backbone by means of one condyle, or knob-like process, instead of by two, as in the Labyrinthodonts and all the amphibia.

Palæontologists have concluded that Labyrinthodonts, Anomodonts, and Monotreme mammals are all descended from some common amphibian stock. If this theory is true, the Anomodonts retained, while they were yet in existence, structures showing a relationship with mammals or a foreshadowing of them, which in all other reptiles have been altogether lost.

Palæontologists are often greatly helped in their studies by comparing extinct types of life with some of the living types which are very old-fashioned and, one would think, *ought* to have become extinct ages ago, but fortunately have not! We mentioned just now the Tuatara, as a case in point; another still more important case is that of the Monotreme mammals—the curious Duckbill and the Spiny Ant-eater (*Echidna*) of Australia (which, of course, is not one of the true Ant-eaters). These two remarkable egg-laying mammals show some wonderful points of resemblance to the Anomodonts—points which cannot, however, be explained without entering too much into dry details of Anatomy. Indeed, they are a great puzzle to naturalists, and one hardly knows whether we ought to call them reptiles or mammals; but since they do suckle their young ones (after a primitive fashion, though), they have been placed with the latter; still it requires a slight stretch of imagination to look upon an egg-laying creature as a mammal.

Leaving now the Anomodonts, we pass on to consider another order—the Rhyncocephalia.

The first fossil Saurian on record, and one which is the only representative of its order, is the *Proterosaurus*[1] *Speneri* of Von Meyer. The generic name records this fact, while that of the species is in honour of its first describer, Spener, a physician at Berlin, in the year 1710. It was found in a slab of "copper-slate," from the Permian beds of Eisenach, in Thuringia. Spener figured the slab showing, chiefly by impressions, the skull, vertebræ, and bones of the fore foot. The original specimen is now in the Royal College of Surgeons, where it forms part of the Hunterian series of fossils. It was obtained from a copper-mine near Eisenbach, at a depth of one hundred feet from the surface. A second specimen, showing the two fore limbs, a hind limb, and part of the trunk, was described by Link in 1718.

[1] Greek—*proteros*, earlier; *sauros*, lizard.

This singular type of reptile, which stands by itself (see Appendix II.), has been studied by Cuvier, Von Meyer, Owen, Huxley, and Seeley, besides Spener. In the eighteenth century very absurd notions were in vogue about fossils. People could not then bring themselves to believe that they were the remains of creatures that once lived, and, to get out of the difficulty, invented the most fanciful theories, invoking the aid of a supposed Plastic Force, which, they said, caused stones to assume fanciful shapes like shells and other living things! But Spener, using his own judgment, set aside these notions, and was sure that his specimen represented what was once a living Saurian. His only doubt was whether the reptile had been a crocodile or a lizard; but he favoured the former view, and tried to speculate how a crocodile could have come into Germany. The great geographical revolutions revealed by geology were then undreamt of. Kundmann of Breslau, in 1737, considered it to be a large-headed lizard. This conclusion was largely adopted by Cuvier, who, in 1818, made the animal universally known as "the fossil Monitor of Thuringia;" but he had never seen a specimen, and so was dependent on the drawings published by Spener and others. The skull is more lizard-like than crocodilian. There are mainly two interpretations of this fossil. (1) Cuvier and Huxley class it unreservedly with the lizards. (2) Von Meyer and Owen consider that it represents a new type of reptile. Von Meyer published an elaborate monograph on the subject, with nine folio plates (1856). Professor Seeley also considers that it shows no particular affinity with any living group of reptiles. It evidently came of a very ancient stock. With regard to its habits, Owen concluded that it was an aquatic animal. He says, "The strength of its neck and head, and the sharpness of its teeth, enabled it to seize and overcome the struggles of the active fishes of the waters which deposited the old Thuringian copper-slates." There can be but little doubt, however, that it was capable of walking on land as well as of disporting itself in the water. Certain ganoid fishes abounded in the seas of the Permian period. These, as Owen points out, may have been its prey.

The beak-headed lizards, or *Rhynchocephalia*,[1] belong to another very remarkable order of reptiles that lived during the New Red Sandstone period. The order is now almost extinct, being only represented by the Tuatara, or *Sphenodon*, of New Zealand.

There are three extinct reptiles to be considered under this head before we conclude our survey of Triassic reptiles, and first we will take the little *Telerpeton*. It was found in the New Red Sandstone strata of Elgin—then considered to be *Old* Red Sandstone, and a great controversy took place with regard to their geological age. But now most geologists accept the view that they are of New Red Sandstone age, as asserted by Professor Huxley, and others. The discovery of *Telerpeton* was an important event in the scientific world at the time, chiefly because the Elgin sandstones were then considered to be of Old Red Sandstone age. The following notice appeared in the *Elgin Courant* of October 10, 1851:—

"GEOLOGICAL DISCOVERY.—A fossil has been obtained from the Old Red Sandstone at Spynie, near Elgin, which serves to establish the fact that air-breathing vertebrata of the order sauria existed during the deposition of the Devonian system of rocks, which hitherto had only been surmised from the discovery of impressions of footprints on the surface of the strata. The fossil, or rather impression, was brought to a gentleman in Elgin, who now possesses it, by a quarryman, who, with praiseworthy care, preserved the fragments without tampering with the impression to any injurious extent, so that its character is unmistakable. It presents the figure of a reptile about four inches in length, showing fragments of teeth, part of the neck, the backbone and ribs, the pelvic bones and hind legs, with part of the tail—the rest of the tail, which seems to have been long, being still unex-

[1] Greek—*rhunchos*, beak; *cephalos*, head.

posed. [Fig. 31 shows the tail as afterwards exposed to view.] The appearance of this small specimen is very striking—the animal matter in its composition having stained the matrix to

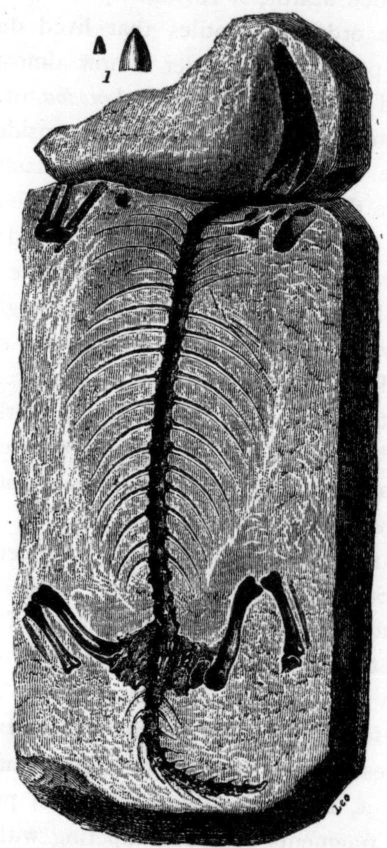

FIG. 31.—Skeleton of *Telerpeton elginense*, from the Elgin Sandstone. Natural size. 1. Tooth. (After Mantell.)

a dark ochre, while the rest of the stone is a pale grey, almost white. No doubt this interesting relic will receive a more scientific description than we can pretend to give, but we cannot resist the impulse to announce the occurrence in this district of an object so well calculated to forward the investigations of geologists."

An accurate drawing of this fossil was sent to Professor Owen by its finder, Mr. Patrick Duff, of Elgin, for his opinion of it. On December 15th, of the same year, Professor Owen's description was published in the same paper, only he called it by a different name, viz. *Leptopleuron*, on account of its slender ribs.

The name now generally accepted denotes the remote antiquity of this reptile.[1]

Dr. Gideon Mantell, the discoverer of *Iguanodon*, described this fossil in a paper read before the Geological Society.[2] A model or cast of this delicate impression is to be seen in the Natural History Museum (Table-case 12, Fossil Reptile Gallery). Unfortunately, the head is not preserved. Dr. Mantell considered *Telerpeton* to be partly amphibian and partly saurian. But, some years later, Professor Huxley came to the rescue, and threw more light on the subject. In a paper read before the Geological Society,[3] he described a new specimen, which showed the skull. This specimen, the property of Mr. James Grant, of Lossiemouth, was sent to him by that veteran geologist, the Rev. Dr. Gordon, of Birnie, by Elgin, who, for fifty years or more, has taken a deep interest in the geology of the district.[4] This specimen is larger, being ten to twelve inches in length. As is usual with fossils from the Elgin sandstone, the bones are represented by casts, but these were well defined. Professor Huxley was able to show that *Telerpeton* has no relationship with amphibians. In all its characters it is a true lizard. Instead of being a low, or "generalised" form, it is somewhat "specialised," or highly developed.

[1] Greek—*tele*, far off; *erpeton*, reptile.

[2] *Quarterly Journal Geological Society*, vol. viii. p. 100 (1852).

[3] Published in the *Journal* of 1867, vol. xxiii. p. 77.

[4] We regret to say that, since these words were written, Dr. Gordon has passed away, at the age of ninety-two. His enthusiasm for science was maintained to the last.

The second example of the reptilian order now under consideration is the *Rhynchosaurus*, discovered by Dr. O. Ward, in a quarry in the New Red Sandstone, at Grinsill, near Shrewsbury. His discovery comprised the skull and a considerable portion of the skeleton (see Fig. 32). Professor Owen also found some remains of the same animal in this quarry.

The animal was about three feet in length. The general aspect of the skull differs from that of modern lizards, and resembles that of a bird or a turtle, which resemblance is increased by the apparent absence of teeth. There was, however, a single row of teeth on the palate. Professor Owen concluded that this ancient creature had its jaws encased in a bony or a horny

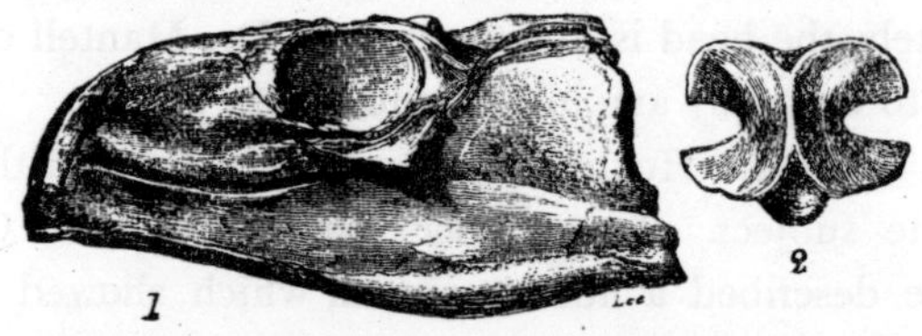

FIG. 32.—1. Skull of *Rhynchosaurus articeps* from the Trias, from the New Red Sandstone near Shrewsbury; ½ natural size. 2. Upper aspect of a dorsal vertebra; natural size. (After Mantell.)

sheath. The fortunate preservation of the skull has brought to light curious modifications of the lizard type of structure, that seem to lead on towards birds and turtles. "The entire reconstruction," says Professor Owen, "of the skeleton of the *Rhynchosaurus* may be ultimately accomplished, if due interest be taken in the collection and preservation of the fossils of the Grinsill quarries."

The *Hyperodapedon* was a much larger creature. It was a terrestrial reptile, some six or seven feet long, and does not appear to have been armed with scutes or spines of any kind. A fine specimen is to be seen at the Natural History Museum (Wall-case 7). Like the *Rhynchosaurus*, it had a broadly triangular form of skull, with the orbits for the eyes directed upwards, and the jaws prolonged into a sharp curved beak, so that it comes into the same order of "beak-headed lizards." But, unlike the latter reptile, it had several rows of well-developed, low, conical teeth, both on the jaws and palate. The specimen at South Kensington shows the head, neck, backbone, and ribs, together with some of the limb-bones. We have therefore ventured on a restoration by Mr. Smit (see Plate IV.), in order to give the reader an idea of its curious head and beaked jaws. (The limbs might be larger.) *Hyperodapedon* must have had a wide geographical range during the period of the Trias; for its remains have also been found in the centre of England (Warwickshire), in Devonshire, and in Central India. A much larger species from India is reckoned to have attained a length of seventeen feet.

Professor Huxley concluded that *Hyperodapedon* and *Rhynchosaurus* do not depart from the ordinary lizard type of structure more than some of our modern lizards do—such as the Monitor, Chameleon, Gecko, and the Sphenodon, to which both are allied. He thinks that, even in Triassic times, the lizard type had become highly specialised.

Ever since the discovery of *Telerpeton elginense*, about forty years ago, the geological age of the Elgin Sandstone has been a subject of keen controversy, and even now there are differences of opinion. Sir Roderick Murchison, late Director of the Geological Survey of the United Kingdom, for a long time held to the opinion that they were of Old Red Sandstone age, as they certainly seem to be if we only consider purely geological evidence; for they rest without any sign of a "break" on undoubted Old Red Sandstone. It was only after Professor Huxley had made his report on *Hyperodapedon* that Murchison began to give way. A little before this he had said, "Let proud Palæontology bow to Field Stratigraphy;" while Huxley said, "Let the battle go as it may, I'll stick to my reptiles!" It was simply a question whether evidence of a purely physical kind (*i.e.* the position of the sandstone and its relation to the strata below, as determined by the field geologist) should give way to evidence derived from Palæontology (*i.e.* evidence from the fossils contained in the sandstone in question). These two methods of determining the age of a rock ought to, and do usually, confirm each other;[1] but here they are at variance. Which, then, are we to believe—the strata, or the fossils? It is an awkward dilemma; but, on the whole, we may say that the presence of reptiles known, from their presence in true Triassic rocks in other regions, to be of Triassic age, gives great weight to the view that the Elgin Sandstones are Triassic, and not of Old Red Sandstone age. The locality became celebrated, and many "Knights of the Hammer" came to see for themselves. The Rev. W. Symonds said he should lose faith in geology if these reptiles turned out to be of Old Red Sandstone age. Professor Huxley compared the battle that raged over this ground to the famous siege of Troy, and said, referring to the views of Murchison and Lyell, that Priam had been fighting on the side of the Greeks, and Nestor on the side of the Trojans! "Under these circumstances," he says, "I thought it best to retire to my tents, and take no part in the fray until my palæontological armoury should yield more efficient weapons. And as my excellent friend, Dr. Gordon, has supplied me with new specimens from time to time, I lived in the hope that one day or other I should be able to make an effective sally."[2] This was in 1858; some years later he was able to convince Sir R. Murchison, and bring him over to the other side.

It only remains, in conclusion, to add that Mr. E. T. Newton, F.G.S., palæontologist to the Geological Survey, has this year (1893) communicated to the Royal Society an important paper on some new reptiles from Elgin.[3] During the last few years a

[1] See *The Autobiography of the Earth*, p. 44.
[2] *Quarterly Journal Geological Society*, xxv. p. 138.
[3] *Philosophical Transactions of the Royal Society*, vol. 184 (1893), B., pp. 431–503.

number of reptilian remains have been obtained from Cuttie's Hill, near Elgin, which are now in the possession of the Elgin Museum and of the Geological Survey. These specimens represent at least eight distinct skeletons, seven of which undoubtedly belong to *Dicynodonts*, and one is a singular horned reptile new to science. All the remains are in the form of hollow moulds in blocks of stone; but Mr. Newton has, with wonderful skill and

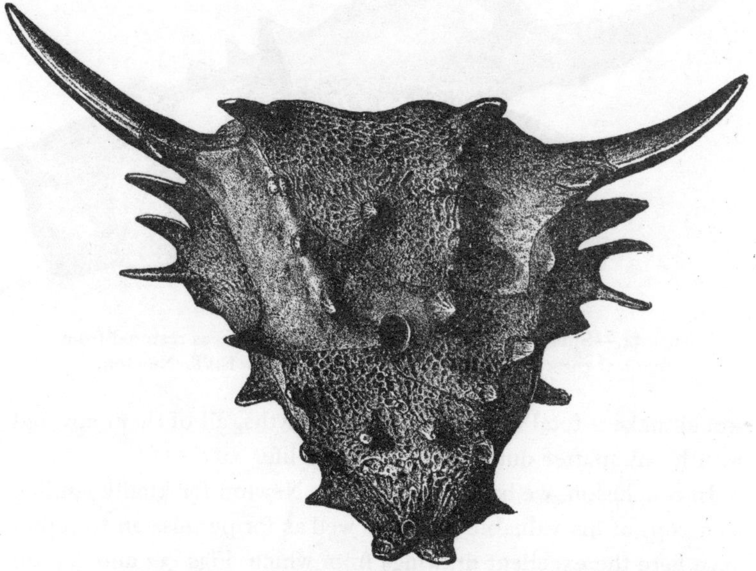

FIG. 33.—Front view of skull of *Elginia mirabilis*, as restored from natural moulds in the Elgin Sandstone, by Mr. E. T. Newton.

patience, made casts in gutta-percha. The casts thus obtained indicate the remains of several species of *Gordonia* (one of the Dicynodonts), and reveal the nature of the skull. *Elginia mirabilis* is the name proposed for the skull of a reptile which, on account of the extreme development of horns and spines, reminds one of the living lizards, Moloch and Phrynosoma. The skull of this ancient saurian (see Figs. 33, 34) is unlike that of any living or fossil form, and seems to show affinities with both the Labyrinthodonts and the lizards. The *Pareiasaurus* seems to be a relation. There are sixteen horns on each side of the skull, varying in length from one-fourth of an inch to nearly three inches, not to mention certain small bosses, which, if included,

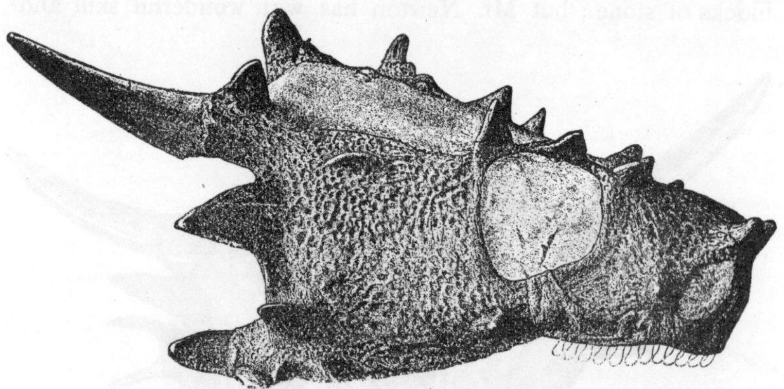

FIG. 34.—Side view of skull of *Elginia mirabilis*, as restored from natural casts in the Elgin Sandstone, by Mr. E. T. Newton.

would make a total of forty bony outgrowths, all of them covered with horny matter during the creature's life.

In conclusion, we have to thank Mr. Newton for kindly sending us a copy of his valuable paper, as well as for permission to reproduce here the excellent drawings from which Figs. 33 and 34 are taken.

CHAPTER V.

THE CROCODILE FAMILY AND ITS HISTORY.

"Geology, of all the Sciences, addresses itself most powerfully to the imagination, and hence one main cause of the interest which it excites."—HUGH MILLER.

THE ancient Egyptians had a great veneration for the crocodile; in fact, they worshipped this monster of their Nile. It was one of the symbols of Typhon, the wicked brother of Osiris, whom they considered to be the cause of every evil. Another of their deities was Sutek, a man with a crocodile's head, and this deity they did all in their power to appease. The palæontologist of to-day also looks upon a crocodile with great veneration; but for very different reasons. To him this voracious reptile is particularly interesting and venerable on account of its long pedigree. The history of the crocodile order, as far as it has yet been traced by geologists, takes us a great way back through "the corridors of Time." We may therefore regard living crocodiles as interesting "survivals" from the wreck of past ages, and cannot but wonder how it is that, while so many other orders of reptilian life have totally disappeared from the theatre of life, this ugly and ferocious reptile should have managed to live on, with comparatively little change either in structure or habits. It is one of Nature's "persistent types"—like the sharks, which have jogged along ever since the Silurian period.

It will be our endeavour, in the present chapter, to give a brief sketch of the geological history of crocodiles. Let it not be supposed that the subject has been fully worked out; such is not the case, but the materials so far collected from the stony record are by no means meagre, and point to certain important conclusions.

Before proceeding on our survey, it may be as well to say a few words about living crocodiles—their structures and habits—so that the reader will be in a better position to appreciate those modifications which we shall find in the fossil forms.

Crocodiles differ in many respects from lizards, with which they were formerly associated on account of their external resemblance. They are thoroughly aquatic in their habits, and the most formidable of all the carnivorous fresh-water animals. Crocodiles and Alligators, when young, and the Gavials throughout life, feed chiefly on fish; but large crocodiles attack every animal which they can overpower. Crocodiles proper are distinguished from Alligators by having the fourth lower tooth passing into a notch on the side of the upper jaw. The teeth are implanted in distinct sockets, while in other recent reptiles they are united to the jaws. The nostrils are situated close together at the upper side of the extremity of the snout; the eyes and ears likewise are near the upper profile of the head, so that the animal can breathe, see, and hear, whilst its body is immersed in the water, the upper part of the head only being raised above the surface. When it goes under water, the nostrils are closed by valves; a transparent membrane is drawn over the eyes; and the ear, which is a horizontal slit, is shut up by a movable projecting flap of the skin. The body is depressed, long, tapering, and protected on the back by rows of solid keeled plates or scutes. The tail, which is long, is powerful, and admirably adapted for propelling the body through the water. The limbs are short but powerful, and the toes are more or less united by a web. There are five digits on the fore limbs, and four on the hinder, and nails are found on three digits fore and aft. The fore limbs are the shortest. The body of the crocodile is carried near the ground in walking, and the hinder part of the belly usually drags, so that the limbs are spread out. The hind limbs have a toothed crest behind, which is formed of about twelve scales. The fore limbs are not used at all in swimming, but are placed flat against the chest. The tail also is protected by a row of scutes. The broad depressed head of the crocodile is very powerful, and the bones of the skull are more consolidated than those of most reptiles. The lower jaw is prolonged behind the base of the skull. The numerous teeth are sharp, conical, and tapering where visible, but they have a hollow, cylindrical fang, which is set in a special hole in the jaw. Crocodiles snap and tear at their prey, and thus wear or drag out their teeth, which are constantly replaced by larger ones, for their first teeth are but small compared to those of old age. Each tooth is hollowed out at the fang, so as to serve for the case or sheath of the germ of the tooth destined to replace it.

Crocodiles inhabit Africa, Southern Asia, the tropical parts of Australia, Central America, and the West Indies. The Indian crocodile is very common in the East Indies and tropical Australia, and has been said to grow to a length of thirty feet; but this must be very exceptional, and eighteen to twenty feet would be an average size.

The following graphic account of the crocodile's habits is taken from one of Mrs. Fisher's (Miss Buckley) delightful works on Natural History.[1] "But it is in the water that we see them in their full strength; there they swim with their webbed feet and strokes of their powerful tail, and feed upon the fishes and water animals—monarchs of all they survey. Nor is the crocodile content with mere fish diet. Often he will lie with his nostrils just above the water, and wait till some animal—it may be a goat, or a hog, or even a good-sized calf—comes to drink, then he will come up slowly towards it, seize it in his formidable jaws, or sometimes strike it with his powerful tail, and drag it

[1] *Winners in Life's Race*, p. 109.

underwater to drown it. For himself, he can shut down his eyelids, and the flaps over his ears, and he has a valve in the back of his throat which he can close, and prevent the water rushing down his open mouth; and after a while he rises slowly till his nostrils are just above the water, and he can breathe freely while his victim is drowning, because his nose-holes are very far back behind the valve. Then, when it is dead, he brings it to shore to tear it to pieces and eat it."

Alligators do not grow to the large size of true crocodiles, and, with the exception of one species lately discovered in China, are found only in America.

The Gharials, or Gavials, may be easily recognized by their long and slender snout. They feed chiefly on fishes, for the capture of which their sharp teeth and slender snouts are well adapted.

Crocodiles of the present day inhabit lakes, rivers, and marshes; but in ancient geological periods they seem to have been less confined to fresh water. The earliest known forms were of a simpler and less "specialised" kind of type. If we had only the existing and more highly developed forms to deal with, there would be little difficulty in separating them from other orders; but in the Triassic rocks we meet with a number of "generalised" forms which seem to be related to the Dinosaurs. For present purposes, and pending further discoveries, the order may be divided as follows. Suborder 1, *Aëtosauria*, represented by the little Triassic *Aëtosaurus*, and another form. Suborder 2, *Parasuchia*, represented by the Triassic *Belodon* and *Parasuchus*, etc. Suborder 3, *Eusuchia*, embracing all those of succeeding periods, including the present. The Right Hon. Prof. Huxley, a good many years ago, made a careful study of extinct crocodiles, and arrived at the conclusion that the oldest crocodiles (*parasuchia* of his classification) differ less than living ones from the Lizards, and that the oldest Dinosaurs approach a less specialised form of Lizard.

One of the oldest fossil crocodilians known is the quaint little *Aëtosaurus*, discovered in the Triassic strata of Würtemburg, which was completely encased in an armour of bony plates (see Plate VI.). Certain bones in its feet, corresponding to those in the soles of our feet, are much elongated, and in many points it seems to approach the carnivorous Dinosaurs, such as Megalosaurus.[1]

Another very ancient crocodile—also from the Trias at Würtemburg—but a larger one, is *Belodon*[2] (or *Phytosaurus*). The entire skeleton of this creature is not known, and there-

FIG. 35.—The oldest fossil Crocodile (*Belodon*), restored, after Fräas, from New Red Sandstone strata, Germany.

fore we cannot state its length; for, as all geologists are aware, the Triassic strata of Europe do not contain well-preserved fossils. A cast of the skull of this important fossil may be seen in the Natural History Museum (Wall-case 2, Gallery 4); also some of its bony scutes. This skull is peculiar in several respects: it is wide at the back, and then becomes long and narrow towards the front part; but in the region of the snout it rises into a long narrow crest or ridge. The teeth are sharp and pointed, with a serrated edge. Fig. 35 is a restoration by Fräas, but incorrect in several respects. We have endeavoured to improve on this in the restoration shown in Plate VI.,

[1] See *Extinct Monsters*, p. 78 (new edit.).
[2] Greek—*belos*, point of a spear; *odous*, *odontos*, tooth.

which is based upon a careful examination of the skull-cast at South Kensington. The scutes in Fräas's restoration are too conical. There are several important points to be noted in this genus, all of which indicate its primitive character compared with later and modern types. One is the nature of the vertebræ, which are hollow or cup-shaped, at both ends (as in fishes, fish-lizards, and most dinosaurs). Each foot was furnished with five digits, whereas modern crocodiles have only four on the hind feet. In all the members of this suborder the peculiar arrangement above referred to, by means of which water is prevented from entering the air-passage when the creature is drowning its prey, is wanting. The palæontologist can trace the gradual evolution of that important modification in forms of succeeding periods; nor is this the only respect in which crocodiles became more highly organised—in other words, more completely adapted to their mode of life and surroundings,—for after tracing their history through the Triassic and Jurassic periods, the palæontologist finds in the next period, viz. the Cretaceous, that they have developed an important modification in their vertebræ, which now show centra hollow in front and convex behind, as in crocodiles of the present day (procœlian group).[1] Another peculiar feature of the skull of *Belodon* is that, unlike all those belonging to crocodiles of succeeding ages, the front pair of nares, or nasal openings, are situated in the middle of the skull, instead of at the end.

Stagonolepis, another Triassic crocodile, belonging to the same primitive suborder, is from the Elgin Sandstone, and deserves to be noticed. This now famous fossil has an interesting history. Our present knowledge of the skeleton is chiefly due to the careful researches of Professor Huxley. In the first instance, only some scutes were discovered; and on these Professor Agassiz founded the genus, thinking they belonged to a

[1] In the earlier forms, the bodies of the vertebræ are cup-shaped, or hollow on *both* sides (amphicœlian).

THE OLDEST KNOWN CROCODILE (BELODON). NEW RED SANDSTONE PERIOD. ALSO THE LITTLE AËTOSAURUS.

PLATE VI.

ganoid fish. To be quite correct, we should say it was only impressions of scutes that Agassiz had to deal with. They looked like scales of the Gar-pike (*Lepidosteus*). "The angular form of these impressions," he said, "allows of no doubt that the fish whence they proceeded was a great ganoid, similar to *Megalichthys*" (a Carboniferous fish). He went on to say that the slab containing these impressions came from Lossiemouth, and was *Old* Red Sandstone. But, as we have already explained (see p. 89), these sandstones have been shown by Professor Huxley to be of *New* Red Sandstone age, although they rest without any break on Old Red Sandstone. Sir Charles Lyell, however, suspected that the scales belonged to a reptile, and expressed his doubts to Hugh Miller, who was fully convinced they belonged to a fish.

In 1858, Sir Roderick Murchison visited the locality. On examining the bony remains found associated with the scutes of *Stagonolepis*, and preserved in the Elgin Museum as well as the collections of Mr. Patrick Duff and the Rev. G. Gordon, he was so impressed with their obviously reptilian character, that he used every exertion to gather all the evidence that could be obtained. But Professor Huxley was the first to prove that the remains really belonged to a reptile, and published the results of his labours in a monograph of the Geological Survey of Great Britain, having been requested by the Director-General to report on the remains found at Elgin. The veteran geologist, the Rev. Dr. Gordon, of Birnie, and Mr. Grant, of Lossiemouth, were instrumental in procuring further specimens of different parts of this ancient crocodile, and those reported on by Professor Huxley consisted of vertebræ, ribs, teeth, part of the skull, some bones of the limbs, and other parts—not enough to furnish a restoration of the skeleton, but enough to give a very fair idea of it. The conclusions arrived at in this report were, that *Stagonolepis* resembled one of the existing Caimans of intertropical America, except that it had a long narrow skull, like that of a gavial. The scutes formed an armour for the under side of the body, as well as for the back, and there were four rows of them along the back, and eight on the under side, whereas modern crocodiles have none on the under side, though some alligators have. The teeth had short, swollen, and rather blunt crowns, like the back teeth of some modern crocodiles. Although the jaws were long, it was more like a modern crocodile than the extinct Teleosaurus to be presently described.

In the same strata were found some footprints; these were also reported on by Professor Huxley, who thought at first that they were probably the tracks of *Stagonolepis;* but he afterwards modified his former opinion, and doubted if they could be referred to any known form of reptile or amphibian. They certainly, however, resemble the tracks of a crocodile in several respects. The tracks are figured in one of Professor Huxley's papers.[1] The length of this ancient crocodile may be put down at from sixteen to eighteen feet.

Certain strata in India, which may be approximately of the same age, have yielded the remains of another type (*Parasuchus*) belonging to the same suborder, but it is very imperfectly known.

We now pass on from the scanty remains of Triassic rocks to consider the abundant examples found in the Jurassic deposits—some of which are complete, and show every bone in its place. These extinct crocodiles, and all the succeeding ones (including modern forms), belong to the third suborder, the Eusuchia, but are divided into two series—amphicœlian and procœlian, according to whether the centra of their vertebræ are hollow on both sides, or hollow in front and convex behind.

There are several families in the first series, of which the Teleosaurs[2] are an example. One of the earliest descriptions of an ancient fossil reptile was that published by Messrs. Wooller and Chapman in the *Philosophical Transactions*, as far back as the year 1758, whose quaint descriptions are accompanied by

[1] *Quarterly Journal Geological Society*, xv. (1859), p. 440.

[2] Greek—*tele*, far away; *sauros*, lizard.

ANCIENT CROCODILIANS. JURASSIC PERIOD.

Pelagosaurus. *Teleosaurus.*

PLATE VII.

a figure of the specimen, which was about nine feet long. It was found near Whitby, on the Yorkshire coast. In a letter to John Fothergill, M.D., Captain W. Chapman says, "The place where these bones lay was frequently covered with sea sand to the depth of two feet, and seldom quite bare, which was the occasion of their being rarely seen; but being informed that they had been discovered by some people two or three years ago, we had one of them with us on the spot, who told us that, when he first saw it, it was intire [*sic*], and had two short legs on that part of the vertebræ wanting towards the head. Although we could not suspect the veracity of this person, we thought he was mistaken, for we had hitherto taken it for a fish. But when we took it up and found the *os femoris* [femur] above mentioned, we had cause to believe his relation true, and to rank this animal among those of the lizard kind; by its length (something more than ten feet) it seems to have been an allegator [*sic*]; but I shall be glad to have thy opinion about it." Subsequent writers entertained various opinions about the fossil. Camper, for example, pronounced it to be a whale, meaning perhaps a dolphin. But, as Cuvier remarked, the presence of teeth in both jaws proves that it could not belong to a whale. Fig. 36 is not a very correct restoration of the genus, especially with regard to the limbs. The representation in Plate VII. will convey a more accurate idea of the animal.

The form of the head approaches that of the gavials, but the beak, or mandible, is more slender, and the teeth more numerous. The first few teeth at the end of the snout are very long. The skin was covered by broader and thicker scutes than those of a crocodile, and these were overlaid in such a manner as to form a strong, flexible coat of mail; they are also deeply pitted. In general shape the *Teleosauri* were more slender than gavials; their feet were better adapted for swimming than for walking, the fore feet being much smaller than the hind ones. It is evident that they could move with great rapidity in the water, but on land their movements were probably more difficult. *Teleosaurus Chapmani* has at least one hundred and forty teeth, while a modern gavial has one hundred and twelve.

A beautiful little specimen of *Teleosaurus*, from the Lias at Monheim, in Franconia, was figured and described by Sommering in the Memoirs of the Academy of Munich in 1814. This example is about three feet long. In the Oolite of Caen, in Normandy, very fine specimens of *Teleosaurus* have been found; and from these, the illustrious Cuvier first determined the character and affinities of the original.

In the year 1791 a second specimen of a long and slender-

FIG. 36.—A Crocodile of the Jurassic period (*Teleosaurus*) restored.

nosed crocodilian was found in the Lias near Whitby, between Staithes and Runswick; and another more perfect skeleton was discovered at Saltwick, near Whitby, in 1824. This was one of the finest *Teleosauri* ever discovered, and is figured by Dean Buckland. It is now preserved in the Whitby Museum, where it was examined by Owen. The snout is lost, but if this be allowed for, the length of the creature would have been quite eighteen feet. The teeth, one hundred and forty in number, are all small, slender, and placed nearly in a straight line. The toes were terminated by long and sharp claws. The backbone is composed of sixty-four biconcave vertebræ, and the largest vertebræ are three inches in length. The genus *Teleosaurus* is only found in the Lower Jurassic strata, and is abundant in the Stonesfield slate of Oxfordshire, and in the nearly equivalent strata of Caen, in Normandy. In the Oxford Clay (see table of strata given in Appendix I.) is found another genus of the same family, known as *Steneosaurus*, but it does not appear to have survived beyond the epoch of the Kimmeridge Clay. Geoffroy St. Hilaire recognised that this later form made a nearer approach to the modern gavial than did the *Teleosaurus*. *Pelagosaurus* [1] is an allied genus represented by two species. The remains of the small *P. typus* are especially abundant in the Upper White Lias of Normandy; and so marvellously are many of the specimens preserved, that every detail of their skeleton is known. In the Natural History Museum may be seen a beautiful reproduction of an entire skeleton of one of these little crocodiles from the Lias of Curcy, Normandy, as prepared by the late Prof. E. Deslongschamps (Gallery No. 4. See Fig. 37). The original specimen consists of seventy-five parts, and is now in the Natural History Museum of Paris. The Professor, with great care and labour, himself worked it out from the matrix. The Bath Museum also contains some excellent specimens.

The abundance of the remains of this family of armoured crocodilians throughout the Liassic and Oolitic formations shows how numerous these carnivorous saurians must have been in the marshes, deltas, and estuaries of the islands and continents of those remote ages. Like the gavials of India, they doubtless swarmed in the rivers and lakes, and preyed on the fishes that lived in the seas of those periods. The crocodile, the gavial, and the alligator all love warmth, and are now found only in hot countries; we may therefore infer that the climate of the period in question was warm. Such an inference is abundantly supported by other kinds of geological evidence, with which we need not trouble the reader.

Metriorhynchus is another genus belonging to the same family, but presenting certain peculiar features such as are unknown in any other members of the whole crocodilian order. It comes

[1] Greek—*pelagos*, sea; *sauros*, lizard.

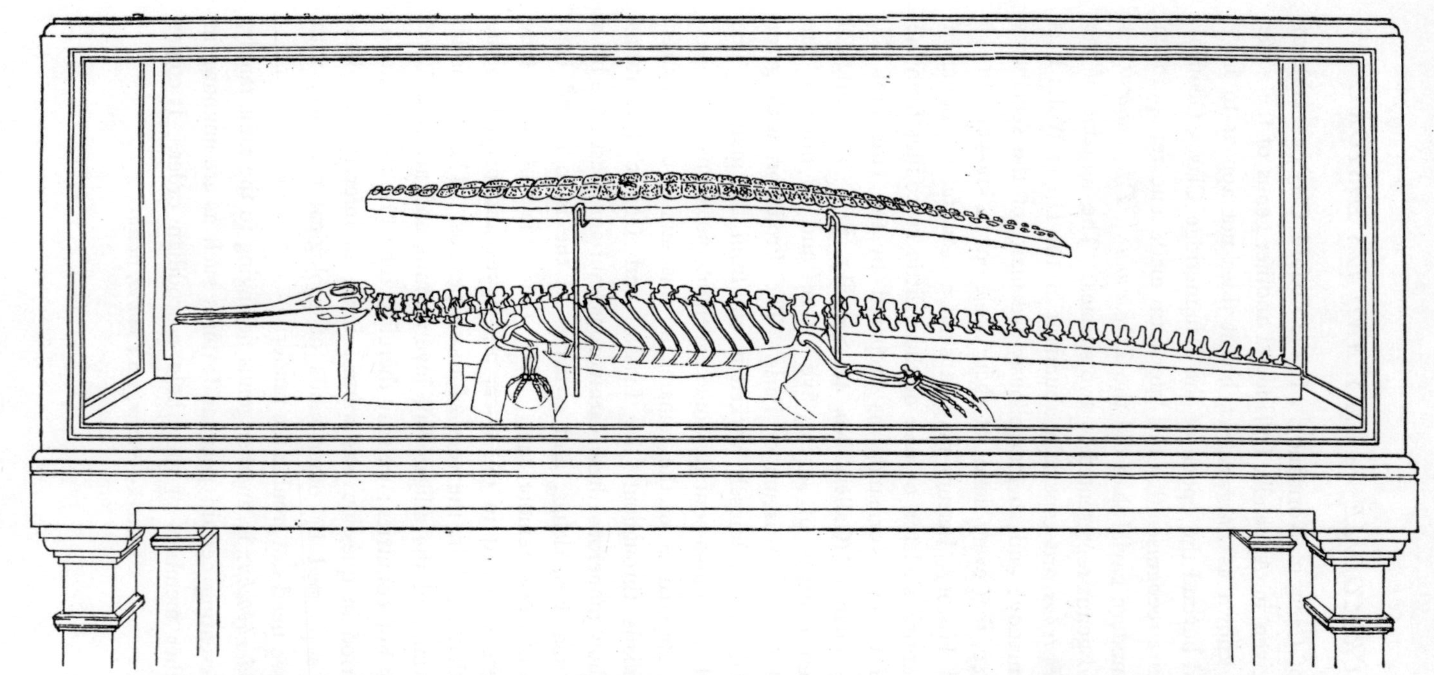

FIG. 37.—Cast of a model of the crocodilian reptile, *Pelagosaurus typus*, from the Upper Lias, Curcy, Normandy. (From a drawing by J. R. Hutchinson.) The original specimen was worked out from the matrix by Prof. E. Eudes Deslongchamps. It consists of seventy-four parts, and is now in the Natural History Museum of Paris. This cast is in the Natural History Museum, London.

from the Kimmeridge Clay of Normandy, and, together with certain other genera, forms a group by itself. In this group we notice an absence of an armour of bony plates or scutes, such as all other crocodiles have. Like the *Ichthyosaurus* described in our former work,[1] it had bony "sclerotic" plates in the eye. Now, it is a remarkable fact that hardly any animal, either living or extinct, is known to have these sclerotic plates and at the same time to have scutes on its body. Some fine examples of complete fossil crocodiles have been obtained from the Kimmeridge Clay near Peterborough, by Mr. Alfred N. Leeds, of Eyebury, whose successful labours as a collector are well known to palæontologists (Wall-case 2, Gallery 4).[2] *Geosaurus*, a more specialised form, from the Kimmeridge Clay of England and corresponding beds on the Continent, was considered by Baron Cuvier to hold an intermediate place between the crocodiles and the monitors. Its eyes were protected by sclerotic plates; the teeth were large and slightly recurved, and the vertebræ biconcave. This genus probably attained a length of ten or twelve feet.

Passing on now to a later age, that of the Chalk (Cretaceous period), we come to *Goniopholis*,[3] an ancient crocodile with teeth, which are cylindrical and smooth at the base, with a somewhat conical crown, the surface of which is strongly marked by numerous ridges. Teeth of this kind are not uncommon in the strata of Tilgate Forest and other localities of the Weald of Sussex, and were first described by Dr. Mantell in 1822. The teeth and bones found there vary considerably in size: some appear to have belonged to individuals not more than eight or ten feet in length; others are twice as large, and indicate crocodiles eighteen or twenty feet long.

[1] *Extinct Monsters*, p. 39 (new edit.).

[2] Three of these have just been placed in the Fossil Reptile Gallery of the Natural History Museum.

[3] So named from the rectangular form of the scutes (Gk. *pholides*).

In the summer of 1835, the workmen employed in a quarry near Swanage, on splitting asunder a large slab of Purbeck limestone, perceiving teeth and portions of bones exposed on the corresponding surfaces of the slabs they had separated, carefully preserved the two pieces of stone, which fortunately were purchased for Dr. Mantell by a friend who happened to be in the neighbourhood. When first received, the slabs only showed obscure indications of the remains that careful chiselling

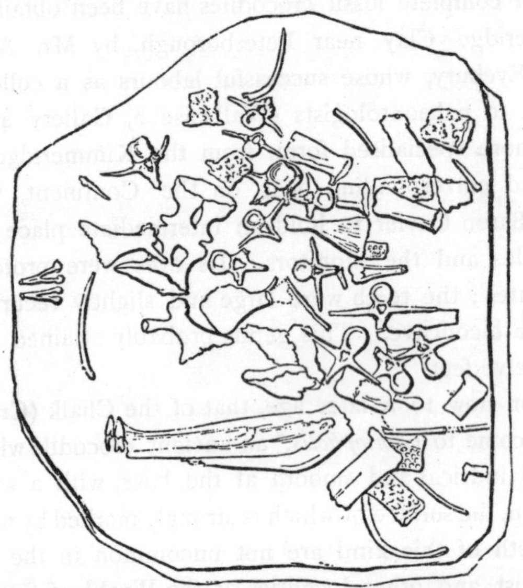

FIG. 38.—Fossil remains of a crocodilian reptile (*Goniopholis*), from the Wealden deposits, Swanage, 1835. (After Mantell.)

subsequently brought to light. After much labour, Dr. Mantell succeeded in developing the detached parts of the skeleton now visible (see Fig. 38). These two slabs, with their corresponding surfaces, may be seen in the Fossil Reptile Gallery, in the Natural History Museum (Wall-case No. 2). In the accompanying diagram (Fig. 38), showing one of the slabs, the reader will easily recognise ribs, vertebræ, scutes, and other parts of this important find. The big bone below is part of the lower jaw containing two teeth. It is not unusual to find specimens of these teeth partly decomposed, and disclosing the young growing tooth inside (see p. 99). The large bony scutes are scattered over the slab, and offer several interesting peculiarities. In the first place, their size is remarkable, some of them being as much as six inches long, and two and a half inches in breadth; their outer surface is impressed by many deep, rounded, or angular pits. In number and size these remind us of the scutes of the Teleosaurs; but they are usually longer in proportion to their breadth. Their most remarkable feature, however, is a kind of "peg and socket" arrangement, by means of which those of the back were fitted together, like tiles on the roof of a house —a structure which is highly characteristic of the large, bony, and enamelled scales of many extinct ganoid fishes (such as *Lepidotus*).

Fragments of the scutes are often found in the Wealden strata, and the earliest specimens collected by Dr. Mantell, from the resemblance of their corrugated surface to that of certain plates in the shield or carapace of the soft-skinned turtles, were figured and described as such in his "Fossils of the Tilgate Forest." It was only when he saw similar scutes associated with crocodilian bones in the slabs from Swanage that he learned their true character. From the size and strength of these bony plates, their highly imbricated surface, and the interlocking arrangement, Professor Owen considered that the *Goniopholis* was better protected than even the extinct *Teleosaurus*.

We may therefore justly conclude, from the skeleton of this ancient crocodile, that it was a powerful carnivorous creature, resembling in its habits the existing gavials and crocodiles, and frequenting the rivers and marshes of the country inhabited by those great terrestrial saurians, such as the *Iguanodon*,[1] with whose remains its bones and teeth are generally found associated throughout the Wealden deposits of England and Germany.

[1] *Extinct Monsters*, p. 98 (new edit.).

Since Mantell's days, further discoveries in Belgium have furnished materials for an almost complete reconstruction of the skeleton of *Goniopholis*. Fig. 39 shows *G. simus* from the Wealden strata of Bernissart (now famous for the discovery of a number of specimens of *Iguanodon*) as mounted and restored by M. L. de Pauw, in the Royal Museum of Natural History at Brussels. Fig. 40 shows an allied genus (*Bernissartia*) treated in the same way. These beautiful specimens, on which much care and labour have been bestowed, show that the under side, as well as the back, was protected by scutes. The two drawings[1] have been made by our artist from photographs, and some of the wires on which the bones were mounted are seen.

In both of these genera the vertebræ are of the primitive type, viz. hollow on both sides, like those of a fish.

Some very large teeth, with highly striated surfaces, and three or four inches in length, are found in strata of the Lower Greensand age and the chalk: to these the name *Polyptychodon*[2] has been given, but whether the creature that owned them was a crocodile or a dinosaur was for some time a matter of doubt, but they are now known to be pliosaurs; we have already pointed out that certain of the more ancient crocodiles show a decided approach to the latter order. Some of these teeth resemble at first sight those of the great sauroid fish *Hypsodon* of Agassiz.

From what has already been said, it will be clear that crocodiles were flourishing in abundance all through the Mesozoic or Secondary Era. It was then an "age of reptiles," and these saurians doubtless only had their share in the good time which the whole class seem to have enjoyed.

Continuing our brief survey of the history of this interesting order, we find, on tracing the fossil forms up to Cainozoic or Tertiary periods, that they were still fairly abundant in Europe and

[1] From the *Bulletin de Musée Royal d'Histoire Naturelle de Belgique*, tom. II (1883). Paper by M. Dollo.

[2] Greek—*polus*, many; *ptuchos*, fold; *odous, odontos*, tooth.

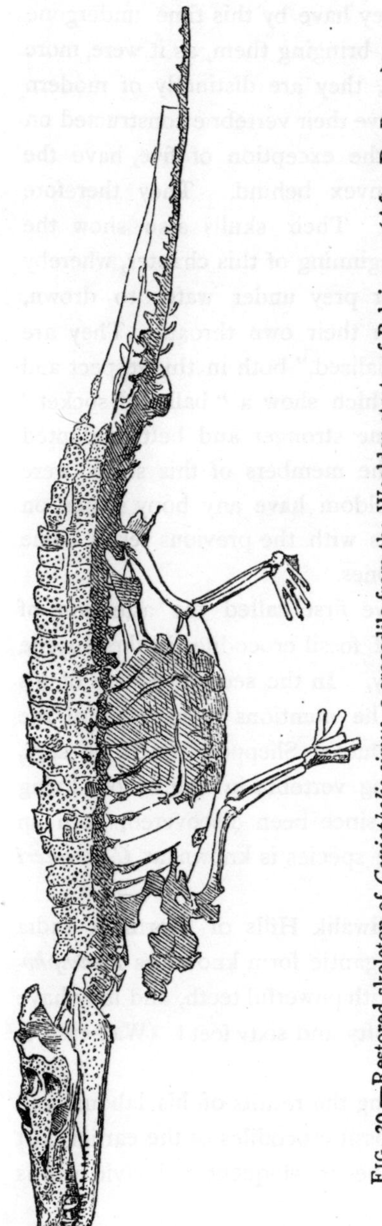

FIG. 39.—Restored skeleton of *Goniopholis simus*, a crocodile from the Wealden strata of Belgium. (After De Pauw.)

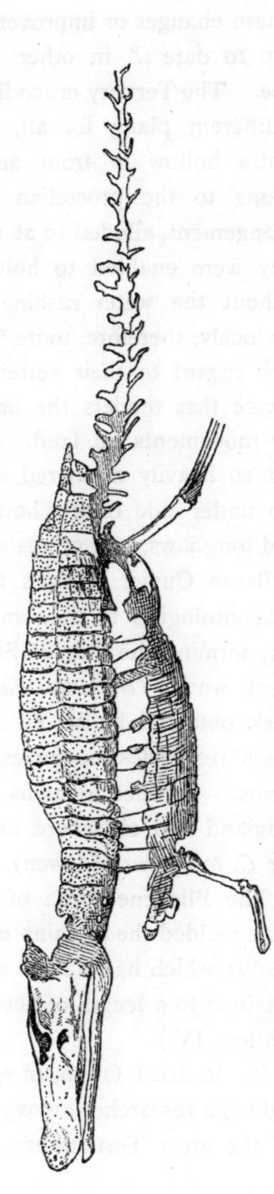

FIG. 40.—Restored skeleton of *Bernissartia Fagesii*, a crocodile from the Wealden strata of Belgium. (After De Pauw.)

other parts of the world. But they have by this time undergone certain changes or improvements, bringing them, as it were, more "up to date;" in other words, they are distinctly of modern type. The Tertiary crocodiles have their vertebræ constructed on a different plan; for all, with the exception of five, have the centra hollow in front and convex behind. They therefore belong to the procœlian series. Their skulls also show the arrangement, alluded to at the beginning of this chapter, whereby they were enabled to hold their prey under water to drown, without the water rushing down their own throats. They are obviously, therefore, more "specialised," both in this respect and with regard to their vertebræ, which show a "ball and socket" device that renders the back-bone stronger and better adapted for movements on land. But the members of this series were not so heavily armoured, and seldom have any bony scutes on the under side of the body. As with the previous series, some had long jaws, and others short ones.

Baron Cuvier appears to have first called the attention of palæontologists to the remains of fossil crocodiles in the Eocene clay forming the Isle of Sheppey. In the second edition of his great work, *Ossemens Fossiles*, he mentions a vertebra of the neck obtained by M. G. A. Deluc at Sheppey, which, he says, much resembles the corresponding vertebra in one of our living forms. Further remains have since been discovered, both in England and elsewhere, and the species is known as *C. spenceri* (or *C. toliapicus* of Owen).

The Pliocene strata of the Siwalik Hills of Northern India have yielded the remains of a gigantic form known as *Rhamphosuchus*, which had a large skull with powerful teeth, and may have attained to a length of between fifty and sixty feet! (Wall-case 2, Gallery IV.).

Sir Richard Owen, in reviewing the results of his labours and valuable researches among the fossil crocodiles of the earlier part of the great Tertiary era, describes in eloquent and vivid words the scenes which presented themselves to his powerful imagination. His words may well be quoted here.

"At the present day, the conditions of earth, air, water, and warmth, which are indispensable to the existence and propagation of these most gigantic of living Saurians, occur only in the tropical or warmer temperate latitudes of the globe. Crocodiles, Gavials, and Alligators now require, in order to put forth in full vigour the powers of their cold-blooded constitution, the stimulus of a large amount of solar heat, with ample verge of watery space for the evolutions which they practise in the capture and disposal of their prey. Marshes with lakes; extensive lakes; large rivers—such as the Gambia and Niger, that traverse the pestilential tracts of Africa, or those that inundate the country through which they run, either periodically, as the Nile for example, or with less regularity, like the Ganges; or which bear a broader current of tepid water along boundless forests and savannahs, like those ploughed in ever-varying channels by the force of the mighty Amazon or Orinoco;—such form the broad features of the destructive existence of the carnivorous and predaceous Crocodilian reptiles. And what, then, must have been the extent and configuration of the Eocene continent, which was drained by the rivers that deposited the masses of clay and sand accumulated in some parts of the London and Hampshire basins to the height of one thousand feet, and forming the graveyard of countless Crocodiles and Gavials? Whither trended that great stream, the haunt once of Alligators, and the resort of tapir-like quadrupeds, the sandy bed of which is now exposed on the upheaved face of Hordwell Cliff?

"Had any of the human kind existed and traversed the land where now the base of Britain rises from the ocean, he might have witnessed the Gavial cleaving the waters of its native river with the velocity of an arrow, and ever and anon rearing its long and slender snout above the waves, and making the banks re-echo with the loud and sharp snappings of its formidable jaws. He might have watched the deadly struggle between the Crocodile and Palæothere, and have been himself warned by the hoarse and deep bellowings of the Alligator from the dangerous vicinity of its retreat. Our fossil evidences supply us with ample materials for this most strange picture of the animal life of ancient Britain; and what adds to the singularity and interest of the restored *tableau vivant* is the fact that it could not now be presented in any part of the world. The same forms of crocodilian reptile, it is true, still exist, but the habitats of the Gavial and the Alligator are wide asunder, thousands of miles of land and ocean intervening: one is peculiar to the tropical rivers of continental Asia, the other is restricted to the warmer latitudes of North and South America; both forms are excluded from Africa, in the rivers of which continent true Crocodiles alone are found. Not one representative of the Crocodilian order naturally exists in any part of Europe; yet every form of the order once flourished in close proximity to each other in a territory which now forms part of England."[1]

[1] *History of British Fossil Reptiles.*

CHAPTER VI.

SOME NEWLY DISCOVERED DINOSAURS.

"DRAGONS OF THE PRIME."

OF the many strange forms of ancient animal life brought to light by the labours of geologists and collectors in various parts of the world, perhaps those of the Dinosaurian order are the most wonderful. Not only in size, but in strangeness and variety, they may be said almost to stand alone. They were indeed the veritable dragons of old time, and seem to come nearer than any other antediluvian animals to the monsters of fairyland, whose acquaintance we all made in the nursery.

In our previous work we devoted several chapters to the consideration of many examples of this remarkable order. It has long been known that these creatures were in existence at a comparatively early period in the world's history, but until the last few years their remains had not been discovered in any rocks older than those of the Triassic or New Red Sandstone period. All the forms which we endeavoured to describe belonged to the Jurassic and Cretaceous periods, and it was only while going to press that we learned, from a paper by Professor Marsh, of the fortunate discovery of a large part of the remains of several Dinosaurs in the well-known Connecticut Sandstone, of Triassic age, in America.

For many years—in fact, ever since the famous Connecticut Sandstone footprints were studied and described by Hitchcock and Deane—geologists have literally been on the track of these earlier and hitherto unknown forms. They could only make rather wild guesses about them from the numerous footprints left behind on those old sands of time, but could not actually hunt them down to their lairs, as we have seen already (chap. i.). Some of these guesses were certainly rather shrewd, while others were far away from the truth. But this was only what might be expected from the nature of the case, for fossil footprints are most difficult things to interpret. We require to know a good deal more of the exact nature of the tracks made now by living animals, under all sorts of conditions, before we can expect to be able to decide, with anything approaching to certainty, the nature of the tracks left by primitive creatures ages and ages ago, and even then we should still be confronted with this grave difficulty—that the creatures that have passed away from the face of the earth were, in many cases, so very different from those with which we are now more or less familiar, that it is no easy matter to decide what kind of impression they would leave behind when "making tracks," as our trans-atlantic cousins say.

However, "the great bone-hunting chief," as the Indians call Professor Marsh, has at last succeeded in hunting down several new Red Sandstone Dinosaurs to their last resting-place, now more than full five fathom deep in the bowels of the earth.

In the present chapter we propose first to give some account, in as brief and simple a manner as possible, of these very interesting and recent discoveries, and then to say a few words about one or two other Dinosaurs, some of small size, others of large dimensions, which were not treated of in our previous work. In this way we hope to complete our account of the Dinosaurs, and bring them, as it were, "up to date."

The newest thing in Dinosaurs seems to be the *Anchisaurus*[1] of Marsh, of which he has made a restoration of the skeleton,

[1] Greek—*anchi*, near; *sauros*, lizard.

as shown in Fig. 41. But it was only by putting together the results of several discoveries that he was enabled to give the complete outline of its skeleton as represented in the accompanying figure, so it may be well to give a brief account of the history of the discoveries that led up to this much-desired result.

As far back as the year 1818, a portion of a skeleton was dis-

FIG. 41.—A carnivorous Dinosaur, *Anchisaurus colurus*, from New Red Sandstone strata, North America. (After Marsh.)

covered in the Connecticut Valley, near Windsor. Another was found near Springfield, and described by Hitchcock, in 1865, under the name *Megadactylus*. Later on, in the year 1884, Professor Marsh announced another discovery, near Manchester, Connecticut, in almost the same geological level, or "horizon,"—

that is to say, in strata of about the same age,—of the remains of an animal of larger size, but in many respects nearly allied to the one described by Hitchcock. Both apparently belong to the same genus. These remains seemed to represent an animal about six to eight feet long; but, unfortunately, they are not complete, although when first discovered the skeleton probably was complete, and, with proper care, might have been preserved entire. Another discovery was reported by Professor Marsh in 1891, when two interesting specimens were secured for the Yale College Museum; one of these was a species of the *Anchisaurus*. Its skull and the greater portion of the skeleton were found in place. The skull is of moderate size and delicate structure, and in general shape somewhat resembles that of the curious New Zealand reptile known as the *Sphenodon*, or Tuatara (see p. 86). The vertebræ and limb-bones are hollow, and the whole skeleton was lightly built; the neck and tail were of moderate length.

Writing in the *American Journal of Science* for February of this year (1893), Professor Marsh reports that the skeletons of five small Dinosaurs have now been discovered in the Connecticut Sandstone. They are sufficiently well preserved to give us most valuable information with regard to all the chief characteristics of the animals to which they once belonged. They were of moderate size, and all were found in about the same geological horizon. From the teeth, as well as from other characters, it is concluded that they were carnivorous, and of this there can be but little doubt. It is equally certain that other large forms of Dinosaurs lived contemporaneously with these; for they have left numerous tell-tale tracks; but no bones have yet been discovered such as might have belonged to them.

With the more complete materials now in his possession Professor Marsh has been enabled to make the restoration of the skeleton of the genus *Anchisaurus*, seen in Fig. 41. We may attribute a length of about six feet to the creature when alive. The skeleton, from which this restoration was chiefly made,

THE OLDEST KNOWN DINOSAUR (ANCHISAURUS). FROM NORTH AMERICA. NEW RED SANDSTONE PERIOD.

PLATE VIII.

was discovered entire, and apparently in the position in which the animal died. Unfortunately, some of the vertebræ belonging to the neck and tail were lost before the importance of the specimen was realised, but the skull and nearly all the rest of the skeleton was saved. In order to complete the outline, Professor Marsh made use of the corresponding parts, fortunately preserved, in another specimen of an allied species, viz. *Anchisaurus solus*, found in the same locality. Although it would doubtless be preferable to make restorations from a single specimen only, because the result would be free from all doubt (except perhaps in some of the positions assigned to limbs, etc.), yet this is seldom possible. Those who make extinct forms of life their study are frequently obliged to adopt the method of scholars in reconstructing the text of some classical author, who, when breaks, or *lacunæ*, occur in one manuscript, consult another one which they consider trustworthy, and from it fill in the gaps.

The beautiful restoration of this Dinosaur shown in Plate VIII. is based on the figure of the skeleton shown above. The *Anchisaurus colurus* was one of the most slender and delicate of all the Dinosaurs yet discovered, being only surpassed in this respect by the little bird-like form, from the Solenhofen slate, known as *Compsognathus*, the skeleton of which is reproduced for comparison on p. 133. This was the creature of which Professor Huxley remarked that no one could look at it without concluding that it must have hopped about on its hind feet, like a bird. The position chosen for the *Anchisaurus*, as represented in our plate, is one which, doubtless, the creature was in the habit of assuming during life. But, at the same time, there can be but little doubt that it was also in the habit of progressing on all-fours. In fact, some persons may think that it looks a little "top-heavy," so to speak, in this semi-erect position assigned to it. But, as other carnivorous Dinosaurs of a later period must have walked erect on their hind legs, we may be allowed to follow Professor Marsh in this matter, and to adopt his conclusion.

It would be needless here to give the various arguments that may be adduced to favour this view, but it may be pointed out, in passing, that such a position is rendered more possible by the way in which the bones of the pelvic region, namely, those known as the ischium and the pubis, are directed, one forward and the other backward. Compare, for example, the same bones in the *Claosaurus* and the *Ceratosaurus*, represented in Figs. 45 and 43, and in Plates IX. and X.

Another little piece of evidence pointing in the same direction is the fact that the fore-fingers terminate in very sharp claws, which seem to be better adapted to the purpose of seizing and holding the prey than for walking on. A prominent and important feature of this Dinosaur is the small head. Notice also its long bird-like neck. The ribs of the trunk, as well as those of the neck, are very slender. From the nature of the vertebræ composing the creature's tail, it is evident that, unlike that of any other Dinosaur, the tail was slender and flexible, as we see in the case of modern lizards. In shape this tail must have been nearly round; here, again, is a point of difference from other Dinosaurs. As we shall see in the case of another animal to be presently described, the dinosaurian tail was usually of a compressed or flattened form, and doubtless was a great help in swimming.

But perhaps the most interesting conclusion connected with this discovery is that the *Anchisaurus* was probably the creature that made some of the tracks found in the strata in which its remains were buried up. For half a century or more these tracks have been a fruitful source of contention, and now at last we seem to be literally "on the track" of one of the many antediluvian creatures that made them. This is very satisfactory, and shows that time solves most things. Professor Marsh points out that, on a firm but moist beach, only three-toed impressions would have been left by the hind feet, and the tail could have been kept free from the ground. On a soft muddy shore, the claw of the first toe of the hind foot would have left its mark, and perhaps the tail also might have touched the ground. As we have already remarked, such additional impressions have been found on the Connecticut Sandstones (see Figs. 3 and 4).

One other carnivorous Dinosaur, named *Ammosarus* by Professor Marsh, and allied to the one above described, is also partly known in the same strata, but we cannot at present give any account of it. One or two other forms from Triassic strata are only imperfectly known; such as the *Thecodontosarus*, from the Triassic conglomerate of Clifton, Bristol, and *Zanclodon*, from Germany and South Africa.

The reader will find in chap. v., p. 101, some account of an interesting fossil crocodile called *Belodon*, that flourished at the same time as these Dinosaurs; but our knowledge of the fauna of this period is unfortunately small, on account of the unfavourable nature of the New Red Sandstone strata, which are too porous to allow of the proper preservation of bones.

Taking one consideration with another, there can be but little doubt that Dinosaurs flourished vigorously during the period of the New Red Sandstone. We have two kinds of evidence to confirm such a conclusion; one is the large number and variety of footprints which they have left behind—although, of course, we cannot suppose that they were all made by Dinosaurs; for, as we have seen, some were probably due to tortoises, while others were probably made by Labyrinthodonts and perhaps even Anomodonts. Another piece of evidence is the known fact that Dinosaurs were abundant in the succeeding Jurassic period. Judging from the size of some of the footprints, it would appear that not a few of the Triassic forms attained large dimensions. And, if the strata had been more favourable, we should have had more of them preserved. But we must pass on to consider a few Dinosaurs of two later periods, some of which have only lately been found, while all of them possess characters of great interest to naturalists and palæontologists.

The first of these is a ferocious-looking beast, recently described by Professor Marsh under the name *Ceratosaurus*. This name, which signified "horned saurian," was given to it on account of the large horn it carried on the head, as shown in Mr. Smit's restoration (Plate IX.).[1] The *Ceratosaurus* is a good typical example of a carnivorous Dinosaur; it was of considerable size, being some twenty-two feet long, and probably a dangerous enemy to some of the herbivorous forms, such as the *Brontosaurus*, one of which

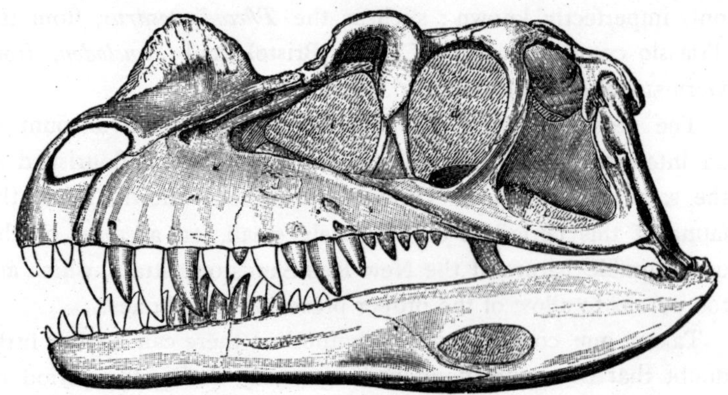

FIG. 42.—Skull of *Ceratosaurus nasicornis*. (After Marsh.)

is represented in our plate as being attacked. This new Dinosaur resembles somewhat the *Megalosaurus*, described in our former work,[2]—so much so, indeed, that Mr. Lydekker considers it should be included under that same genus, which has been rendered classic by the labours of Professors Buckland, Phillips, and Owen. The large bones of the limbs of these formidable flesh-eating monsters were hollow, and many of the vertebræ are deeply excavated, except at the ends. A study of the skeleton of this creature shows clearly that it was constructed with a view to

[1] Greek—*ceras*, horn; *sauros*, lizard.

[2] *Extinct Monsters*, new edit., p. 76.

lightness and strength combined. The skull of this species, *C. nasicornis* (see Fig. 42), is very large in proportion to the rest of the skeleton, and its general structure is light and open. The nasal bones support a large horn-core; the side of this elevation is very rough, and marked with what are called "vascular grooves," all of which tends to confirm the idea that it supported a high horn, which must have been a most powerful weapon of offence and defence.

FIG. 43.—Skeleton of a large carnivorous Dinosaur, *Ceratosaurus nasicornis*, from Jurassic strata, North America. (After Marsh.)

The jaw-bones are large and massive, each being provided with fifteen teeth, in use, which are large and trenchant, indicating clearly the ferocious character of the animal. The eyes seem to have been protected by partially overhanging protuberances, shown in the skull (see Fig. 42). The tail was long and well adapted for swimming. The fore limbs are very small; each has four fingers, all armed with sharp claws, and it certainly looks as if they could have been but little used for supporting the body,

PLATE IX. A LARGE HORNED DINOSAUR (CERATOSAURUS) FROM NORTH AMERICA. JURASSIC PERIOD.

but for holding the prey they would be very useful. The bones, of the pelvic region (that of the hind limbs), are particularly interesting. In the first place, the three bones of each side, instead of being more or less independent of each other, are fused together. The two long bones, known as the ischium and the pubis, were each expanded at their ends into a kind of foot; and Professor Marsh concludes that, when the creature sat down, it was partly supported by these protuberances, and he thinks that some of the Connecticut impressions prove this point.

Just as the *Ceratosaurus* of the New World is represented by, or rather corresponds to, the *Megalosaurus* of the Old World, so the other newly discovered Dinosaur, which we are about to describe, shows a good deal of correspondence with the now well-known *Iguanodon.* It is called by Professor Marsh, *Claosaurus;* and it represents a whole family of herbivorous Dinosaurs. In length it was nearly thirty feet, and must have stood nearly fifteen feet high. The remains of this remarkable creature were discovered by two of Professor Marsh's assistants in the Laramie beds of Wyoming, associated with those of the still more strange *Triceratops*,[1] and also of some diminutive mammals lately discovered.

The most important feature in the skeleton of the *Claosaurus* is the skull. This is long and narrow, and, as seen in Fig. 44, shows a large muzzle, the surface of which is considerably roughened, showing that it was covered, during life, with hard and thickened skin. The brain of the creature was very small in proportion to its skull. The orbits for the eyes are large, suggesting large eyes; and its sense of smell was probably pretty keen. The teeth resemble those of *Hadrosaurus*, an allied form.[2] Succulent vegetation probably was the main diet of this animal. The whole back-bone, or vertebral column, was found complete, with the exception of a few little vertebræ from the end of the tail. Altogether there were about ninety vertebræ in the back-bone of this

[1] *Extinct Monsters*, frontispiece. [2] See p. 138.

powerful creature. The reader will perceive, from the drawing of the skeleton, in Fig. 45, that this Dinosaur must have possessed a long and powerful tail, such as would have constituted a most useful organ of propulsion through the water. As in its ally, *Iguanodon*, the vertebræ of the tail are compressed in a vertical

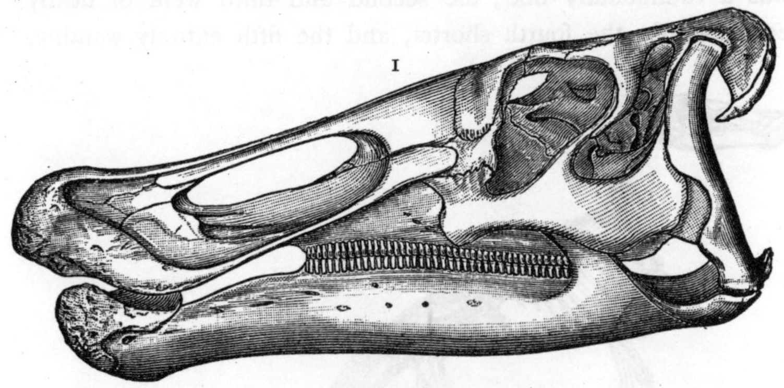

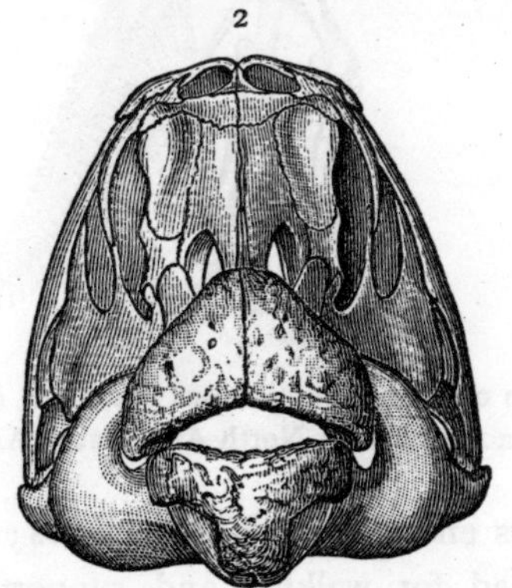

FIG. 44.—1. Skull of *Claosaurus annectens*, side view. 2. The same, front view. (After Marsh.)

direction. Plate X., which is a restoration, should be compared with the drawing of the skeleton in Fig. 45. Another important feature of the skeleton is the remarkable ossified tendons, by means of which the neural spines of the vertebræ were held together, and so strengthened. They increase in number in the region of the thighs, and also in the first part of the tail.

The reader will have noticed the smallness of the fore limbs compared with the hind limbs. This is often the case with Dinosaurs. Only three fingers of the hand were used—the first was a rudimentary one; the second and third were of nearly equal length, the fourth shorter, and the fifth entirely wanting.

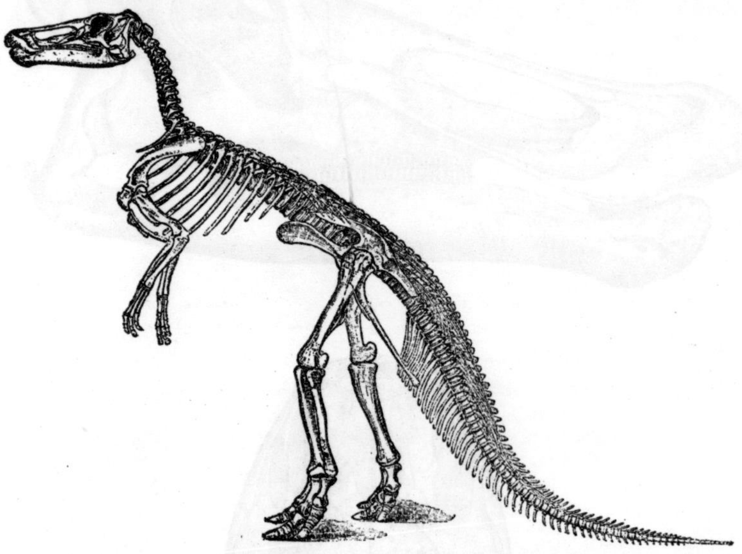

FIG. 45.—Skeleton of a large herbivorous Dinosaur, *Claosaurus annectens*, from Cretaceous strata, North America. (After Marsh.)

These long fingers ended in hoofs, not claws; the fore limb as a whole was adapted for walking and support, rather than for purposes of prehension, although the latter might have been expected from its small size and position. The hind limb has three digits, all well developed and massive. These were covered with fairly broad hoofs, and it is clear that this Dinosaur was in the habit of walking on its hind legs. It is distinguished from *Hadrosaurus*, a near ally, by the fact that its limb-bones, instead

A LARGE HERBIVOROUS DINOSAUR (CLAOSAURUS) FROM NORTH AMERICA. CRETACEOUS PERIOD.

Plate X. The bird represented on the right is a young *Hesperornis*.

of being hollow, were all solid. This piece of evidence tends to confirm the idea that *Claosaurus* was fond of the water; for it has been pointed out by Owen, Marsh, and other authorities, that land animals, having to support the weight of their bodies on land, have hollow bones; while those which live in the water, like whales, being buoyed up by the action of the water, and thus having so much weight taken off their hands and feet, have their limb-bones solid.

The late Sir Richard Owen, in his great work on *British Fossil Reptiles*, made some valuable remarks on the probable habits of Dinosaurs, which deserve careful attention. He points out that actual observation of a swimming crocodile testifies to the fact that the fore-limbs are laid flat and motionless upon the sides of the chest. Their chief swimming organ is, of course, the tail, which is both long and powerful, and, in some forms, is flattened in a vertical direction. Most of the Dinosaurs at present known have the crocodilian swimming organ; and some, such as the *Claosaurus* and the *Iguanodon*, reveal it considerably compressed vertically. In those forms it must have been a cumbrous impediment in the way of walking on land; but in others, such as *Anchisaurus*, the tail was rounder and more lizard-like, and doubtless these carnivorous forms were less hampered in their movements on land. Concerning the smallness of the fore limbs, Professor Owen suggests that the forward dash of the creature through the water in pursuit of fishy prey would be facilitated by this reduction in size of the fore limbs, which would take no part in movement, unless in the way of steering. On land they might be of some use in scratching out the nest for the eggs, as well as in helping to support the body.

With regard to the relations of this Dinosaur to others, Professor Marsh considers that it represents a separate family. The reasons for this opinion, being based on certain peculiar features which require technical description, need not be given here. The result of a scientific examination of the skeleton of

Claosaurus shows it to be a highly specialised—that it is, a highly developed—animal, making allowance, of course, for the fact that it was after all only a reptile; or perhaps it would be better to say a Dinosaur, for the Dinosaurs, as an order, were very unlike any reptiles now living. We are so apt to judge of extinct reptiles by those orders which have survived, that a Dinosaur appears as a very different thing, to which the term reptile seems hardly appropriate. But mere questions of classification need not trouble us much; for even with regard to living animals there is considerable difficulty. But when naturalists come to try and make out a scheme of classes and orders for extinct animals, they are beset with far greater difficulties. All the existing classifications of animals, whether living or extinct, must be regarded as purely temporary arrangements.

The points of resemblance between birds and Dinosaurs have been pointed out by Professor Huxley. This resemblance shows itself most strikingly in the hind limbs; and we have already alluded to the fact that for a long time the Connecticut footprints (see chap. i.) were taken for those of birds. Considering their very bird-like character, such a conclusion was very natural; and even now, it would not be safe for geologists to assume that there were no representatives of the bird class at the time when those footprints were made.

It has sometimes been asserted that birds were derived from Dinosaurs in the course of evolution, but this is not a correct statement of the views of leading naturalists, such as Professor Huxley. They would only venture to suggest that probably both were derived from some common ancestor, which clearly is not the same thing. This view has been pretty widely accepted by students of ancient animal life—partly, perhaps, in deference to so great an authority as Professor Huxley; but is it not just possible that these points of resemblance (which are too obvious to be denied) may be due to the simple fact that, for reasons best known to themselves, many

Dinosaurs walked, in a more or less erect position, on their hind limbs? Similar habits, if continued for age after age, must tend to produce somewhat similar appearances, at least externally. Take, for example, the case of whales and fishes; the original land mammal from which whales are descended has, in the course of time, become so fish-like in appearance that even in these modern days there are some who yet speak of them as fishes! The shape of a whale is fish-like; it has lost its hind limbs through disuse; it has changed its fore limbs into paddles, which have a certain fin-like aspect; and its cousin, the porpoise, has developed a big triangular fin on the back! To take another example: Pterodactyls are reptiles which acquired the power of flight; and some, at least, of them must have borne a resemblance to modern bats.[1] And yet pterodactyls and bats are in no way related. The pterodactyl was a lizard that could fly; while the bat is a mammal that can do the same. There is no question of relationship. Now let us apply the same kind of reasoning to Dinosaurs and birds. Palæontologists tell us they are related: the question is, are the resemblances of their skeletons (in certain points) to birds to be taken as implying relationship however distant? For our part, we confess to being not quite convinced. Dinosaurs walked, ran, and perhaps some of the smaller ones even hopped on their hind legs; would it not follow as a matter of course that certain rather bird-like developments would follow? Let us reason the matter out. For a horizontal body, like that of a lizard, to be properly poised on two legs instead of four, the weight of the viscera must be transferred backward, and the forward, or anterior part of the body lightened. The lower bones of the region of the thighs (pelvis), with the contained organs, are thrown backward, while the fore part of the body and the fore limbs are lightened and much reduced in proportionate size.

Changes such as these might be supposed capable of producing

[1] See *Extinct Monsters*, p. 131 (new edit.).

those bird-like features in Dinosaurs which we have already noticed, and on the strength of which some authorities believe birds and Dinosaurs to have had a common ancestry. We insert here an illustration (see Fig. 46) in which are shown, side by side, the limb-bones and those of the region of the pelvis in a crocodile, a dinosaur, and a bird. (The middle drawing is not

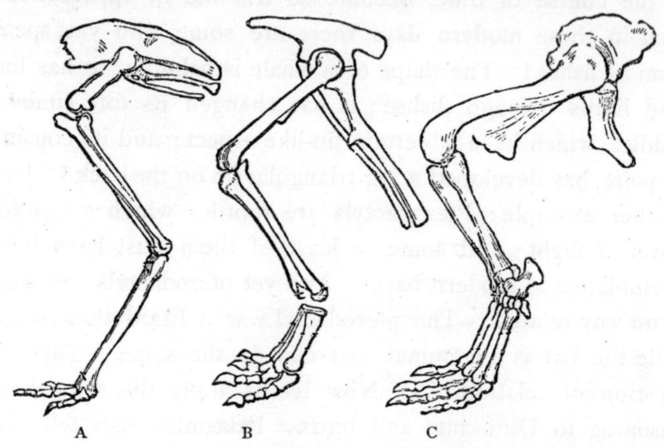

FIG. 46.—Hind limbs of (A) Bird, (B) Dinosaur, and (C) Crocodile.

complete.) Professor Marsh has described some remains of certain small and slender little Dinosaurs (*Hallopus*[1]) about the size of a fowl, which, from the construction of their feet, he believes were capable of jumping.

Mr. J. W. Hulke, another English authority on Dinosaurs, described more than ten years ago a very remarkable and slender form of Dinosaur, on which we must now, in conclusion, say a few words. Professor Huxley has also helped to unravel the meaning of its structure. *Hypsilophodon* is the name this little creature has received, on account of the nature of its teeth. As far as we know at present, it was the smallest of the Dinosaurs with the exception of the *Compsognathus*[2] and *Hallopus*.

[1] Greek—*hallomai*, to jump; *pous*, foot.
[2] Greek—*kompsos*, elegant; *gnathos*, jaw.

It is certainly the least specialised—that is, the least highly organised. It lived during the Wealden period, or, in other words, during the closing scenes of the great Mesozoic era, and so was

FIG. 47.—Skeleton of a small Dinosaur, *Compsognathus longipes*. (From the Solenhofen limestone.)

For restoration, see Frontispiece.

contemporary with others of a much more specialised character. As a rule the remains of animals found in our English Wealden and

Cretaceous strata are always in a fragmentary condition; but in this case we have, fortunately, an exception. Almost the whole of its skeleton is now known, thanks to the labours of Mr. Hulke and the Rev. William Fox. The remains occur in a bed which crops out a short distance west of Barns High Cliff, and passes under the shore a few yards west of Cowleaze Chine, on the south coast of the Isle of Wight. In the first case, it was wrongly described by Sir R. Owen as a young *Iguanodon*. A restoration of its skeleton, according to Hulke, is seen in Fig. 48, and in Plate XI. we have the creature restored to life. The skull shows a combination of characters belonging to a crocodile and a lizard, but is on the whole more lizard-like. The creature was

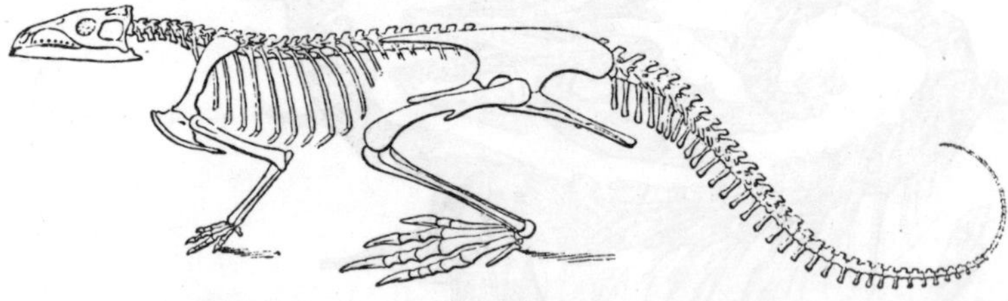

FIG. 48.—Skeleton of a small Dinosaur, *Hypsilophodon Foxii*, from Wealden strata, Isle of Wight. Restored by Professor J. W. Hulke, F.R.S. From the *Philosophical Transactions of the Royal Society*, vol. 173.

about four feet long. In form, mode of attachment, the teeth resemble those of lizards; their crowns are finely serrated, and resemble in miniature the serrations of the teeth of an Iguanodon. The hind limbs are considerably longer than the fore limbs; and the hind foot had four long toes, with a rudiment of a fifth. The sharp pointed and curved claws seem to indicate that it was a rock-climber; perhaps it also could climb trees. The great depth of the tail suggests that it was also a good swimmer. The fore limbs were provided with four fingers and a rudiment of a fifth. The eyes of this creature, which were probably large, contained, or rather were protected by, a ring of bony plates, "sclerotic plates," not shown in Plate XI., which are so well seen in

Plate XI. A SMALL DINOSAUR (HYPSILOPHODON). PTERODACTYLS IN THE AIR. CRETACEOUS PERIOD.

the *Ichthyosaurus*, and in the eyes of some modern birds. This interesting peculiarity is also exhibited by the labyrinthodonts, which probably inherited it from their ancestors, the fishes. The presence of this structure in the *Hypsilophodon* is quite in harmony

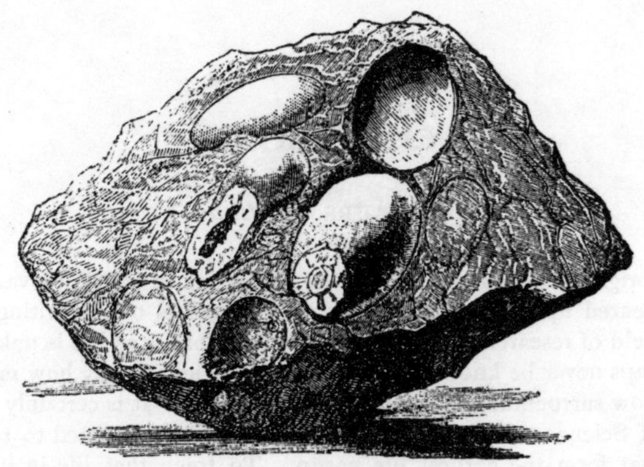

FIG. 49.—A piece of Great Colite limestone with imbedded group of eggs of reptiles. (After Buckman).

with what was said about its being an unspecialised or lowly form of reptile.

Geologists and naturalists may confidently look forward during the next ten or twenty years to the discovery of many new and interesting forms of Dinosaurian life.

CHAPTER VII.

DINOSAURS (*continued*).

"The origin of life, and the order of succession in which its various forms have appeared upon the earth, offer to Science its most inviting and most difficult field of research. Although the primal origin of life is unknown, and may perhaps never be known, yet no one has a right to say how much of the mystery now surrounding it Science cannot remove. It is certainly within the domain of Science to determine when the earth was first fitted to receive life, and in what form the earliest life began. To trace that life in its manifold changes through past ages to the present is a more difficult task, but one from which modern Science does not shrink."—PROFESSOR O. C. MARSH.

WE now pass on to the consideration of some American Dinosaurs, and of Mr. Waterhouse Hawkins's work in restoring extinct animals.

Having undertaken the important task of constructing large models for the Central Park at New York, it became necessary for Mr. Hawkins to make a careful and complete survey of the collections scattered throughout the country. Two months were spent in examining the Museums at Washington, New Brunswick, New Haven, and Philadelphia. Every facility was offered to him at these various institutions; each contained valuable material for the purpose in view, and especially that of New Brunswick. In Philadelphia, at the Academy of Natural Sciences, he met with a rich storehouse of fossil treasures such as were especially needed for illustrating the gigantic forms of life that once peopled the continent of North America.

As had already been pointed out by Dr. Leidy, this collection

Dryptosaurus (*Lælaps*). *Hadrosaurus.* *Great Irish Deer.*

FIG. 50.—Mr. Waterhouse Hawkins's workshop.

proved that the old American Dinosaurs were related to those which had long been known in Europe. The two forms of ancient life first selected by Mr. Hawkins were *Hadrosaurus* and *Dryptosaurus* (*Lælaps*); careful drawings and descriptions of all the fragments of the fossil remains of these two Dinosaurs were made. Fortunately it so happened at this particular time that the curators of the Academy were occupied in extricating from the matrix a valuable specimen of the former Dinosaur recently received from New Jersey. This difficult work was entrusted to Mr. Hawkins. The fossil remains amounted to about one-third of the actual skeleton of the animal; a cast of each bone was made in plaster of Paris, and the missing parts were restored so as to harmonise with the rest of the skeleton. These parts were also moulded, and in this way a complete representation of the bony framework of this extinct monster was obtained, such as is shown in Fig. 50, which shows that its proportions were by no means insignificant.

Hadrosaurus was a near relative of the *Iguanodon*, described in our previous work,[1] so that the one throws light upon the other. In view of the interest attaching to this singular and even grotesque Dinosaur, it may be worth while to give some account of the discovery of its remains. The story may be briefly told as follows:—

Mr. W. P. Foulke, a member of the Academy of Philadelphia, while passing the season at Haddonfield, five miles from Philadelphia, learned that one of his neighbours, while digging marl upon his farm, about twenty years ago, had found some bones. These were described as being of large size and very numerous. Unfortunately, visitors had been permitted to carry away fossils, but those taken were chiefly vertebræ, and it was hoped that, if the old site could be found—for it was quite filled up, and overgrown with shrubs and grass—the rest of the skeleton might be obtained. Permission was given to dig in any part of the farm, and a party of marl-diggers was set to work. The old pit was thus discovered, and, after a time, the diggers were fortunate

[1] *Extinct Monsters*, p. 97, new edit.

enough to come upon a pile of bones, which proved to belong to the creature whose remains were so much desired. With great care the bones were all separately taken out, and transferred to the residence of Mr. Foulke. Dr. Leidy and Mr. Isaac Lea were informed of this important discovery, and they promptly visited the excavation. The bones collected during the time that the excavation lasted were put in order and described by Professor Joseph Leidy, and the animal was named by him *Hadrosaurus Foulkii.*[1] When the spoils were reckoned up, it was found that they consisted of twenty-eight vertebræ, an arm-bone, two bones of the fore-arm, several bones of the pelvis, a thigh-bone, a shin-bone, a fibula, and several bones of the hind foot. The collection also contained nine teeth, and a fragment of a lower jaw. It will thus be seen that all the principal parts of the creature's skeleton were represented; so that Mr. Hawkins was enabled, without drawing too much upon the imagination, to restore those parts that were missing. The restored skeleton of *Hadrosaurus*, as seen in Fig. 50, measured twenty-six feet in length; and stands, on its hinder extremities and tail, thirteen feet, three inches high. These measurements show that the creature must have rivalled in size even the huge Iguanodonts discovered at Bernissart, in Belgium, and described in our previous work.[2]

Without going too much into detail, we may point out a few conclusions to be drawn from these bones. With regard to the vertebræ, or bones of the spinal column, they gradually increase in size from the head to the region of the thighs (pelvis) and beginning of the tail, where they are of the greatest bulk. In the tail they gradually diminish in size without diminishing in length. An important feature is the length of the spinous processes of the caudal, or tail vertebræ; and the well-marked abutments below indicate the existence of well-developed chevron, or V-like bones. These caudal vertebræ, by their number, size,

[1] Greek—*hadros*, powerful; *sauros*, lizard.

[2] *Extinct Monsters*, p. 98 (new edit.).

Hadrosaurus. *Dryptosaurus.* *Plesiosaurus.* *Megatherium.* *Glyptodon.* *Mammoth.*

FIG. 51.—Models by Waterhouse Hawkins, formerly in the Central Park, New York.

and other features, indicate a tail of immense strength, great length, and depth. The bones of the fore limb are entire, and the humerus, or arm-bone, is twenty-two and a half inches in length. This limb was freely movable in all directions. The bones of the hind limbs greatly exceed in size those of the fore limbs. The femur, or thigh-bone, is forty-one and a half inches in length; the tibia, or shin-bone, is over three feet in length, and nearly a foot in breadth at the upper end. These bones of the hind limb were hollow, thus affording both lightness and strength. Professor Leidy inferred, from the manifest disproportion between the fore and the hind limbs, that the *Hadrosaurus* was in the habit of assuming a position like that of a kangaroo, sustaining itself like a tripod, on its enormous hinder extremities and powerful tail.

Such an animal must have been equally at home both in the water and on land: with its powerful tail it was probably a good swimmer, and with its long hind legs it would be able to reach the leaves of trees and shrubs which to less-favoured creatures must have been out of reach.

And here it may be pointed out that these conclusions have been rendered more likely, if not actually confirmed, by the observations of Hitchcock and Deane, alluded to in our first chapter; for they have discovered the impressions of the tails of certain Dinosaurs, made as they walked along or stopped now and then to rest (see p. 19).

Mr. Waterhouse Hawkins's restorations in the Central Park naturally attracted a good deal of attention; and although, in the light of later discoveries, they could certainly be improved upon, yet his endeavours were praiseworthy. The Commissioners for the Park, in their thirteenth annual report, allude to them as follows:—

" A very wide interest, both in this country and in Europe, has been excited among scientific men by this interesting and novel undertaking. The proceedings of the Commissioners of the Park in this matter have been alluded to, commented upon, and commended by scientific journals both at home and abroad. It would be difficult to insure too great care in the preservation of the wonderful remains of animal organisations of past times that are from time to time discovered in different parts of the country. There are examples of fossil remains lying in public and private collections of the country, that, in the interests of science, should be utilised and placed where they can be got at by those specially interested in this department of inquiry. It is very difficult, except through the offer of a reward in money, to impress upon those who, in excavation, casually come upon fossil remains, the importance of handling them with care; they are often, to them, nothing but old bones, and a stroke of the pick or a scoop of the shovel may, in an instant, irrecoverably destroy or cast away a fragment that might serve to establish or refute received ideas of the past eras of our globe."[1]

The *Hadrosaurus*, like the European *Iguanodon*, was a vegetable feeder; the other Dinosaur, *Dryptosaurus* (*Lælaps*), represented in Fig. 51, was a huge carnivorous creature, nearly allied to the *Megalosaurus*, described in our previous work.

We are sorry to hear that these models and others, on which Mr. Waterhouse Hawkins spent so much care, have been destroyed and ignominiously cast into the lake! Incredible as it may seem, we are informed that this wanton piece of destruction was wrought at the bidding of a certain Mayor of New York, who, unfortunately for the cause of Science, considered that the results arrived at by geologists did not harmonise with revealed religion. Let us hope that by this time people have learned the truth so well expressed by the great naturalist Agassiz, who said, speaking

[1] The author is indebted to Professor Rupert Jones, F.R.S., for the loan of the *Twelfth and Thirteenth Annual Reports of the Board of Commissioners for the Central Park, New York*, which are not to be found even in the British Museum Library. The above information is chiefly derived from these Reports, and Mr. Smit's careful drawings (Figs. 50 and 51), are made from two plates therein.

of living animals: "These are but the thoughts of the Almighty uttered in material forms."

Fig. 52 is from a sketch made in the grounds of the Crystal Palace, Sydenham, and shows a large number of the models there set up by Mr. Waterhouse Hawkins. These, although some of them were made from very imperfect materials, such as did not really warrant any attempt at restoration, nevertheless are interesting, historically, as being the first attempts of the kind, and at least serve to draw the attention of those who visit the Palace to the wonders of geology. It will be seen that the huge *Iguanodon* on the left, and the ponderous *Megalosaurus* next to it are very far from the truth, but some of the others further on are much more in accordance with modern ideas on the subject.

A Perilous Fossil Hunt.

We have already alluded, in our previous work, to the dangers and difficulties encountered by Professor Marsh in his explorations, during which he has crossed the Rocky Mountains no less than twenty-seven times. But, in order more fully to illustrate this branch of the subject, it may be worth while to give some further account of these labours, undertaken in the cause of Science. The following story is based upon a narrative which appeared in the *New York Tribune*, December 22, 1874.

In October of that year news reached Professor Marsh that a most promising locality of fossils had been discovered in the "Bad Lands" of Dakota, south of the Black Hills; the character and condition of these bones, as they lay embedded in position, would throw additional light on a problem which he was then trying to work out. It was important to explore this region as soon as possible; for the Indians have a way of carrying a fossil tooth or bone as an amulet or charm—or, as they phrase it, "medicine." The discovery of fossil remains in the locality which Professor Marsh explored was originally made by an Indian,

Iguanodon. *Megalosaurus.* *Pterodactyls.* *Teleosaurus.* *Plesiosaurus.* *Labyrinthodon.*

FIG. 52.—Mr. Waterhouse Hawkins's Models of Extinct Animals in the grounds of the Crystal Palace, Sydenham. From a sketch by J. R. Hutchinson (by permission).

who brought into the camp the molar tooth of an animal which Professor Marsh has named *Brontotherium* (see Fig. 63). The finder carried the tooth in his tobacco-pouch; his notion about it was that it had belonged to "a big horse struck by lightning."

The weather was intensely cold, the season rapidly advancing, and the Indians feverishly sensitive about the approach of the white men. But, great as were the perils, the attractions were greater. This expedition differed from all previous ones in the fact that the Professor, fearing special danger and hardships, took with him no party from New Haven. He depended for assistance in the field on a number of frontier-men, who had been in his employ as collectors and guides in previous expeditions, and on whom he knew he could implicitly rely. Among these was Hank Clifford, who had been his chief guide in the Niobrara expedition in the previous year, and whose knowledge of the country and of the Indians had been fully tested. Other less famous, but promising, aspirants for honours upon the "bone-fields" were attached to the expedition.

Leaving the railroad at Cheyenne, Professor Marsh reached Fort Laramie in the early part of November, and thence proceeded to the Red Cloud Indian Agency, where he concentrated the men and materials of the expedition. The outfit on such occasion includes a great variety of articles—implements of war, of science, and of the kitchen, with abundant means for packing the specimens obtained so that they shall not be injured by the roughest kind of transportation. General L. P. Bradley, Colonel Stanton, Captain Mix, and Lieutenant Hay, officers of the United States army, were of the party that went from Laramie to the Red Cloud Agency; the escort was M. Company of the 2nd Cavalry, Captain John Mix in command. Major A. S. Burt and Lieutenant W. L. Carpenter joined the expedition at the Agency, and greatly contributed to its success.

Unfortunately the bone-field lay north of the White River, in the Indian reservation; and it was therefore advisable to obtain the assent of the Indians, in accordance with a certain treaty. Professor Marsh was anxious to have a willing assent from them, a fight with Indians being, as the newspaper writer remarks, no part of the programme. Shortly after the arrival of the party at the Agency, Red Cloud, the great Sioux chief, put in an appearance, and was welcomed to dinner. The proposed expedition was only partially discussed at the dinner, and Red Cloud's sentiments with regard to it were judged to be not unfriendly. The conversation took place through the medium of an interpreter.

But there were many circumstances making the time unpropitious. Certain recent affairs had very nearly led to bloodshed. The general danger was greatly increased at this time by the presence in the neighbourhood of an extraordinary number of Indians, gathered to obtain their annuities from the Government. Outlaws, renegades, and "Bad Indians" swelled the numbers that surrounded the Agency, and made the neighbourhood unquiet, not to say dangerous. In short, the whole vicinity was alive with Indians, their families, and their ponies. The agent at the post recommended that a guard should be selected from the warriors to accompany the expedition, and very soon assembled a council of leading chiefs to discuss the matter. It soon became evident that the Indians mistrusted the intentions of the bone-hunters, as stated by the agent. White Tail, one of the principal chiefs, at once sprang to his feet and harangued the audience, declaring that the proposed bone-seeking was merely a ruse to begin digging for gold and invading the Black Hills region. The other chiefs applauded with guttural ejaculations of "How! How!" But a conciliatory speech from the Professor, promising that their just complaints should be heard at Washington, and holding out the prospect of pay for Indian services in bone-hunting, turned the scale once more. In this way consent was given for the expedition to proceed, but coupled with an agreement to take a selected guard of young warriors.

Nominally this guard was for protection; but, in reality, the object of the Indians was to keep watch on the proceedings of the bone-hunters. Sitting Bull, one of the most influential chieftains, was to go at their head. Professor Marsh was to let him know when he was ready to move forward.

But the next three days snow fell, which unfortunately caused a delay; for the Indians, having meanwhile got their annuities, changed their attitude and became decidedly aggressive. After much discussion among themselves, they all came to the conclusion that the pretence of seeking fossils was much "too thin." The chief of the bone-hunters was certainly in search of gold. Quite unaware of the change of feeling on the part of the Indians, the party went on with their preparations, and, on the second morning after the issue of annuities, broke camp and proceeded to the Agency, expecting to get the Indian guard. To reach the Agency they had to pass between several villages composed of Indian lodges. The sight of the soldiers and wagons excited the Indians; they gathered in great numbers about the Agency. Red Cloud, when Professor Marsh applied to him, said his young men promised as a guard refused to go, believing the object of the search was gold, not bones. Suddenly a cry of warning was given, and the women and children fled; guns were pointed at the party on every side, and in all directions riders were seen galloping off to the villages and calling together the warriors. The Indians outnumbered the expedition by at least thirty to one. A single shot or the order "forward" would have brought down their fire.

To push on under such circumstances would have been madness. There was but one thing to do, and that was to retire. Reluctantly Professor Marsh gave the order. This movement was the occasion of much jeering on the part of the Indians, and insults were heaped on the party, but, bad as were these insults, they were preferable to bullets. The rest of the day was spent in consultation. On the following day, beef was issued by the agent to the bands entitled to it. Meanwhile, as the result of many consultations, two conclusions were arrived at: (1) that something must be done; (2) that a feast given by Professor Marsh, and a few presents to leading chiefs, were the most promising means of attaining consent.

Accordingly, on the following day, the feast was given; only the more eminent chiefs were admitted, and every detail of Indian etiquette was strictly observed. At its close, after many speeches, Professor Marsh stated the object and character of the expedition, and a reluctant consent was given, with the warning that the Minneconjous were likely to kill the Professor if he crossed the White River. A band of scouts was promised.

Fearing that a consent coupled with so much hesitation might prove unavailable, the Professor resolved to test it. He sent word quietly, late at night after the feast, to his interpreters and guides, to be ready the next morning. But they all alike refused to go. Not a little disappointed at this, the Professor determined to give the Indians the slip. The next night, shortly after midnight, he carried out his intention. Marching down between the villages as silently as possible, the expedition sought the White River at the only spot where, for many miles, it is fordable. The Indian dogs barked furiously as the party defiled between the lodges, but fortunately their owners slept. If the expedition had been attacked at this time their case would have been hopeless. It was a bitterly cold night. The stolen march was soon discovered, but, by means of a forced march, they were able to reach the region they were making for. On arrival, a position of great natural strength was chosen for a camp by one of the military officers. The party was soon able to commence the search for fossils, although watched by Indians at a distance.

The next day the party began systematic work; in fact, the weather was so intensely cold that work became a necessity. As fast as the fossils were secured, they were heaped together, and piles of stones were placed to mark the localities of the

PROF. O. C. MARSH,
Yale Univ., U.S.A.

PROF. E. D. COPE,
Pennsylvania Univ., U.S.A.

THE RIGHT HON. T. H. HUXLEY,
Late Prof. Royal School of Mines.

PROF. KARL ZITTEL,
Munich Univ.

PROF. A. GAUDRY,
Nat. Hist. Museum, Paris.

PLATE XII.

bones in the event of a snowstorm. The cold was intense, at least 20° Fahr. below zero. Four officers and many soldiers were severely frost-bitten; but none of those who were at work digging for bones. At length the cold moderated and there came a snow-storm. The places marked by piles of rock were the scene of renewed labours. Brooms made of bushes and grass were employed at these points to brush away the snow. Meanwhile, in spite of the cold, the Indians had kept their mounted sentinels on the neighbouring hills, watching the operations of the party. Occasionally, in the day time, a few Sioux dropped in with proffers of friendship, probably to obtain a nearer look at the work of the expedition. When success was nearly assured, a more serious cause for alarm was found in the warning given by a party of Indians, who had ascertained that the Northern Indians were coming to make an attack on the camp. To throw the specimens into the waggons and rattle off with them unpacked, was simply to break them in pieces. To pack them at night, burning lights in the tents, would be to invite attack. Great as was the risk of remaining, Professor Marsh, after due consultation with the officers in command of the escort, decided to stay long enough to pack properly. The expedition broke camp the next day, and not a day too soon; subsequent reports state that a large party of Northern Indians scoured the Bad Lands on the following day, in a vain search for "the bone-hunting chief" and his band, who were then *en route* for New Haven.

On his return to the Agency, Red Cloud was among the first to welcome Professor Marsh. Some of the chiefs, to show their good will, proposed to give a dog-feast in his honour, the tender canine being considered by them a great delicacy; but Professor Marsh sent his regrets, and pleaded a previous engagement.

Now that we know some of the results of this most important expedition, it is not to be denied that the dangers encountered were fully justified. About two tons of fossils were secured, and, among them, some very valuable specimens were obtained, especially the bones of certain extinct Miocene mammals. Among other treasures, Professor Marsh thus secured several specimens of a great quadruped known to palæontologists as the *Brontotherium*. (See p. 192.)

CHAPTER VIII.

ANCIENT BIRDS.

"Nor is the value of the doctrine of Evolution to the philosophic thinker diminished by the fact that it applies the same method to the living and the non-living world; and embraces in one stupendous analogy, the growth of a solar system from molecular chaos, the shaping of the earth from the nebulous cubhood of its youth, through innumerable changes and immeasurable ages, to its present form, and the development of a living being from the shapeless mass of protoplasm we term a germ."—The RT. HON. PROF. HUXLEY.

FOSSIL birds are but rarely met with in the stratified rocks; hence our knowledge of the bird life of former ages is comparatively slight. But this is only what might have been expected; for it must be remembered that birds are protected by their powers of flight from perishing in such ways as other animals frequently do. And even should they die on the water, their bodies are not likely to be submerged; for, being light and feathery, they do not sink, but continue floating until the body rots away, or is devoured by some creature, such as a hungry pike.

When did the bird make its first appearance, or *début*, on the earth? in other words, when did that primitive, but as yet unknown reptile from which the feathered tribe came, first take to itself feathers and assume both the habits and appearance of a bird? This is one of those interesting questions which remain to be solved by the labours of the palæontologist—or, more probably, of a generation of palæontologists. We have already alluded (see p. 131) to Professor Huxley's theory that birds are descended from Dinosaurs; but though there is much to be said in favour of the idea, we prefer, for our part, to wait and see what evidence may yet turn up on this subject, and, like the Irishman, to "prophecy *after* the event," *i.e.* when further discoveries have been made. Sir R. Owen never favoured the theory, and, for all palæontologists can tell, it may just as well be that birds and pterodactyls (flying reptiles) both are descended from a common stock; the one line choosing to fly by means of a thin membrane attached chiefly to a single long finger, while the others thought they could do quite as well—in fact better—by growing feathers on their arms and fingers. All great problems in Nature are solved slowly, by the patient accumulation of evidence; and the one above alluded to is no exception to the rule.

Palæontologists are not without evidence bearing on this subject; so perhaps we cannot do better than state briefly what that evidence is, in order to show how near, or how far, we are from a final answer to our question. Though no one can yet say what the very first bird-type was like, we can, at all events, describe the oldest known fossil bird. This is the famous *Archæopteryx.*[1]

Time was—and that within the memory of living geologists—when no fossil birds were known in rocks older than the Tertiary deposits; but the discovery of *Archæopteryx* has changed all that, and we now trace back the bird line to the middle of the great Secondary or Mesozoic Era. This bird was found in the Solenhofen limestone of Bavaria, which is supposed to represent the lower part of our English Kimmeridge clay. First, only the impression of a single feather was known, to which the late Professor H. von Meyer gave the name *Archæopteryx lithographica* (because this stone is much used by lithographers). Later on came the discovery of a magnificent and almost

[1] Greek—*Archios*, ancient; *pterux*, wing.

complete specimen, with beautiful impressions of the feathers. This was named by Owen *A. macrura*, on account of its long tail. The specimen is to be seen in the Natural History

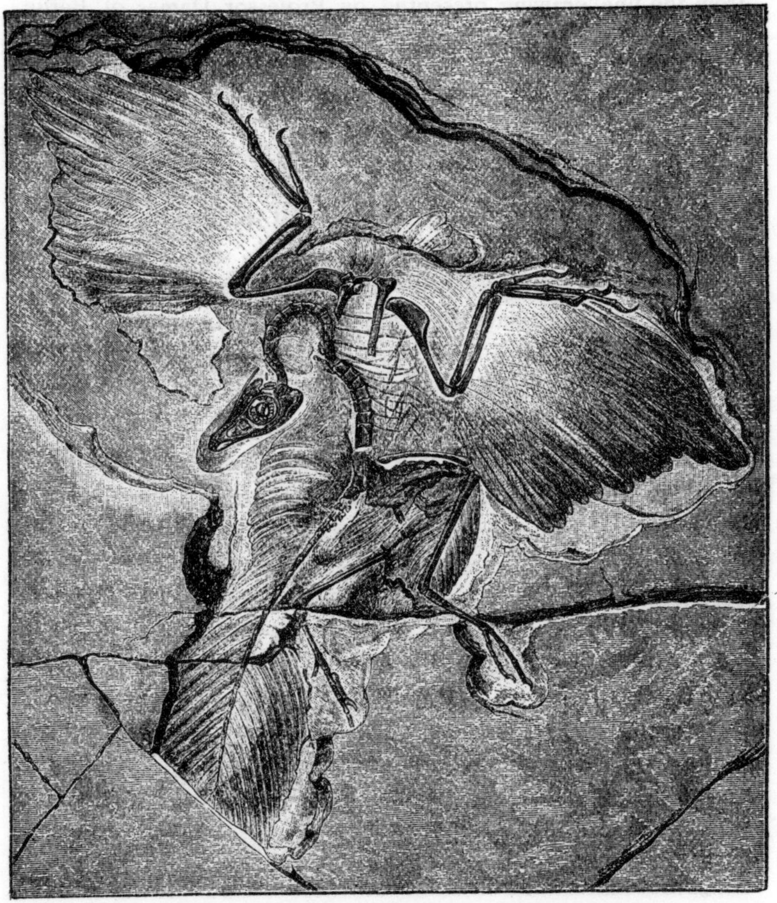

FIG. 53.—The Berlin *Archæopteryx*. (After Dames.)
From the Solenhofen Limestone (see p. 155).

Museum, Gallery No. 11; it was only acquired for the nation by the strenuous efforts and liberality of the late Sir R. Owen (see preface). More recently a still finer specimen, with the head complete, has turned up, and this is now in the Berlin Museum (see Fig. 53). A very good idea of it may be gathered from the fine drawing presented by Professor Dames of Berlin, which is to be seen at the Natural History Museum. Some of those who first studied the creature's anatomy took it for a reptile; and they may well have been puzzled by its curiously "mixed" characters. It is certainly one of the most anomalous types of ancient life that has ever been discovered.

Palæontologists are now agreed that it *was* a bird; for it had feathers and claws, as birds have. But what can we say to a bird with teeth in its jaws, and with a long, lizard-like tail such as reptiles have now, except that this tail was provided with feathers attached in a very peculiar way, contrary to all bird's tails of the present day? A naturalist acquainted only with the avian life of to-day, would certainly say that it upsets nearly all his ideas of what a bird *ought* to be; but that only shows how useless it is (as we have previously pointed out, see p. 73) to lay down rules for Nature. Its vertebræ were bi-concave, like those of fishes and some extinct Saurians. Another very reptilian feature is the presence of "sclerotic plates" in the eye. A pair of feathers sprang from each joint in the tail, which is quite a different arrangement to that in the tail of living birds. The leg-bone and foot are similar to that of a modern perching bird, but then we have seen that some of the Dinosaurs, such as the little *Compsognathus*, had very bird-like feet. The wing shows three free digits or fingers, and it is possible, as Dr. C. H. Hurst[1] has endeavoured to prove, that a fourth finger of greater length and size supported the primary feathers. In form and position these three finger bones are just what may be seen in some young birds of to-day. It has been claimed by some that the fore limb of *Archæopteryx* is reptilian. Another important point, established by Marsh, is that the bones of the pelvis are

[1] *Natural Science*, vol. iii., p. 275.

separate—not united, as in modern birds. In the London specimen (where the skull is lost) there remains a cast of the brain cavity. This reveals the fact that the brain, although small, was more bird-like than reptilian. Were it not for the feathers, perhaps no one would at first have thought of calling it a bird. It combines the features, or characters, of birds and reptiles in a most remarkable way. In size it was about as large as a rook. Are we to call it a Pterodactyl with feathers? That would be to dub it a reptile; and since the leading authorities agree in calling it a bird, we must abide by their decision. Names, however, are sometimes misleading; the main fact we have to remember is, that in former ages some classes of animals which are now sharply marked off from each other were by no means so separated.

A good deal of discussion has lately arisen with regard to the true interpretation of the Berlin specimen, as far as the wing is concerned. The question is, how the feathers were attached; and there can be but little doubt, that in this respect the drawing from which Fig. 53 is taken is wrong. Look at the impressions of the long "primary" feathers, and you will see that at first they curve towards the three fingers, but then turn downwards and bend towards the two bones of the fore-arm. This is due to a mistake, and Dr. Hurst has shown that they probably were attached to a longer and stronger finger which is concealed in the slab below. The first curve of the feathers, as seen in the left-hand corner, should have been continued. But in other respects the Fig. 53 gives a fair idea of the Berlin specimen. Those three slender fingers were probably used for climbing trees, but they are too weak to be of much service in supporting the strong primary quills during flight. In the specimen these fingers lie not in the wing at all, but upon its feather-clad surface. Unfortunately our artist made the beautiful restoration seen in the Frontispiece before the publication of Dr. Hurst's paper; but, with a true naturalist's instinct, he has improved on the drawing seen in Fig. 53 in two respects; first, by showing a distinction between primary and secondary feathers, and, secondly, by not making all the feathers curve backwards, in order that they should appear to have been attached to the ulna (one of the bones of the fore-arm).

If Dr. Hurst's conclusions with regard to the wing of *Archæopteryx* are true, they certainly tend to connect birds more with Pterodactyls than with Dinosaurs, and so militate against the generally accepted view. One cannot help wondering whether this very ancient bird could sing; but, although doubtless the gift of song was not in those Jurassic days so marvellously perfected as it is now, yet we would fain believe that *Archæopteryx* at least *tried* to sing (as suggested in the restoration). If, as seems probable, it lived among trees, one would think that some means of communication with its fellows was almost a necessity.

The next evidences of former bird life are met with in the Cretaceous rocks, and here we find the *Hesperornis*. This bird was a gigantic diver, and its length from the point of the bill to the end of the toes must have been between five and six feet (see Plate XIV.). Its habits are clearly indicated by the skeletons described by Professor Marsh in his splendid monograph on *The Extinct Toothed Birds of North America*. No living birds possess teeth in their jaws, so that the presence of such in *Hesperornis* and other birds of the Cretaceous period at once separates them from those of the present day. It cannot be doubted that this antique diver was carnivorous; it probably devoured fishes. The teeth were set in a groove, and old ones were replaced by young ones growing up from inside the fang. The breast-bone (sternum) was entirely without a keel. The single thin wing-bone (humerus) indicates that its wings were "rudimentary," and quite useless either for flying or swimming (see Fig. 54). Modern penguins use their wings with great effect while swimming under water, but the *Hesperornis* was compensated for the want of wings by its broad tail, which was much expanded horizontally, and doubtless served as an organ of propulsion in diving. (The tail in our restoration is too small.)

The brain was diminutive, and very like that of a reptile. It is impossible, in the absence of feathers, to say exactly what appearance the skin of this bird presented, but in the restoration,

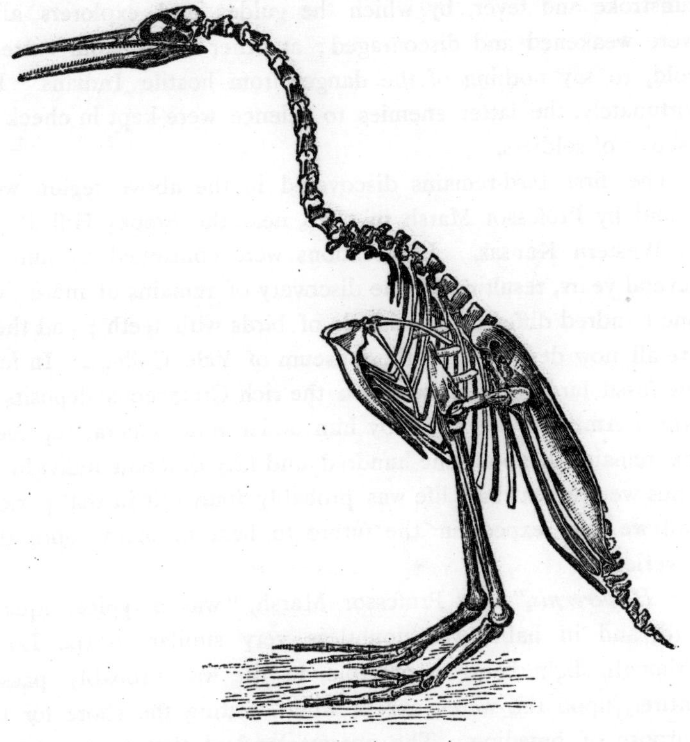

FIG. 54.—Skeleton of *Hesperornis regalis*, from Cretaceous strata, North America. (After Marsh.)

Plate XIV., our artist has to some extent taken a modern diver as his model. The skeleton of this bird presents several interesting points of resemblance with reptiles ; but it will not be necessary to enter into anatomical details here. The remains were discovered by Professor Marsh, in certain marine Cretaceous strata along the eastern slope of the Rocky Mountains, which have also yielded mosasauroid reptiles, Pliosaurs, and Pterodactyls. The valuable specimens on which Professor Marsh's descriptions are based were not procured without considerable trouble and hardship. His exploring parties in the plains of Kansas and Colorado suffered, at one time, from extreme heat, causing sunstroke and fever, by which the guides and explorers alike were weakened and discouraged; at other times, from extreme cold, to say nothing of the danger from hostile Indians. But fortunately, the latter enemies to science were kept in check by escorts of soldiers.

The first bird-remains discovered in the above region were found by Professor Marsh in 1870, near the Smoky Hill River, in Western Kansas. Explorations were continued by him for several years, resulting in the discovery of remains of more than one hundred different individuals of birds with teeth; and these are all now deposited in the museum of Yale College. In fact, the fossil birds now known from the rich Cretaceous deposits of North America, described by him under nine genera, represent the remains of about one hundred and fifty different individuals. Thus we see that bird life was probably abundant in that period, and we may expect in the future to hear of many more discoveries.

"*Hesperornis*," says Professor Marsh, "was a typical aquatic bird, and in habit was doubtless very similar to the Loon, although, flight being impossible, its life was probably passed entirely upon the water, except when visiting the shore for the purpose of breeding. The nearest land at that time was the succession of low islands which marked the position of the present Rocky Mountains. In the shallow tropical sea, extending from this land five hundred miles or more to the eastward, and to unknown limits north and south, there was the greatest abundance and variety of fishes, and these doubtless constituted the main food of the present species. *Hesperornis*, as we have

A GIGANTIC DIVER (HESPERORNIS REGALIS). FROM NORTH AMERICA. CRETACEOUS PERIOD.

PLATE XIV. Length 6 feet.

seen, was an admirable diver; while the long neck, with its capabilities of rapid flexure, and the long slender jaws armed with sharp recurved teeth, formed together a perfect instrument for the capture and retention of the most agile fish. As the lower jaws were united in front only by cartilage, as in the serpents, and had on each side a joint which admitted of some motion, the power of swallowing was doubtless equal to almost any emergency."

It may fairly be concluded that, for a long period of time, circumstances were eminently favourable to *Hesperornis* and its allies; for apparently it had no enemies in the air above, and an abundance of food in the water. We may well imagine that it was more than a match for the gigantic toothless Pterodactyls (such as *Pteranodon*)[1] which hovered over the waters here in such great numbers, and the other inhabitants of the air all appear to have been small. The ocean in which this bird swam teemed with fishes of many kinds, and thus a great variety of food, easily obtainable, was at hand. In this aquatic paradise *Hesperornis* flourished, disturbed only by the serpentine Mosasaurus, which may have been the cause of its extermination. Another bird discovered by Professor Marsh in the same region, also with strangely blended characters, is the *Ichthyornis* (Fig. 55). Unlike the big diver above described, it had well-developed wings and a strongly keeled sternum for the attachment of muscles with which to work its wing. It was about the size of a rock-pigeon. The jaws were armed with teeth placed in distinct sockets, as in some extinct reptiles. The wing bones show that it possessed considerable powers of flight. Here we may note that the Cretaceous birds at present known (some twenty species or more) were apparently all aquatic forms, which, of course, are most likely to be preserved in marine deposits, while the Jurassic *Archæopteryx* was a land bird.

Remains of Cretaceous birds were first found in the Upper

[1] See *Extinct Monsters*, p. 129 (new edit.).

Greensand of Cambridge, and on these bones the genus *Enaliornis* has been founded. In its head and neck it resembled the divers.

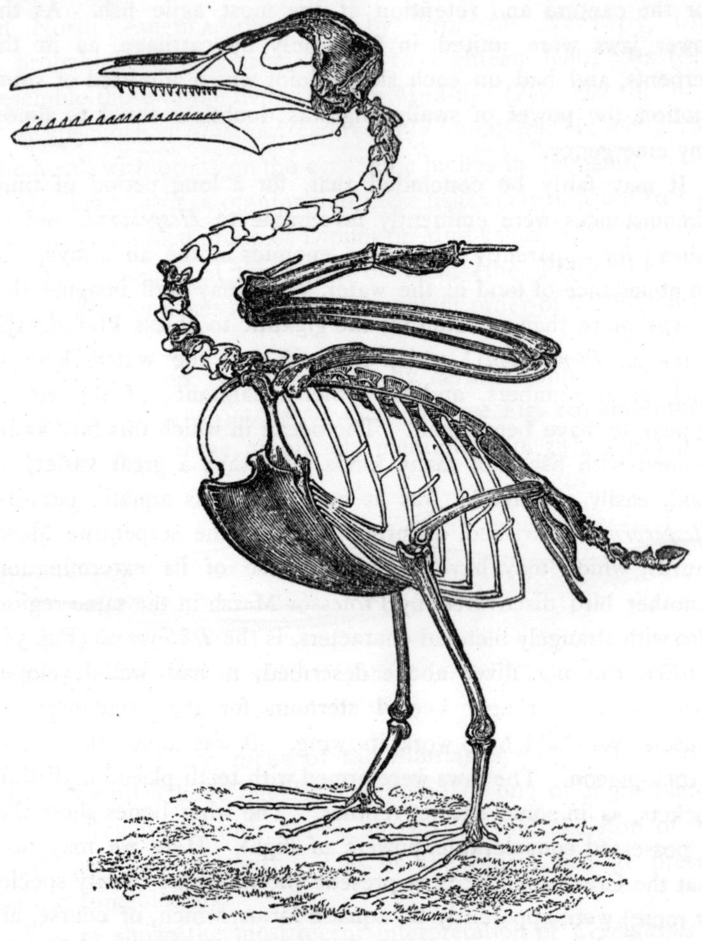

FIG. 55.—Skeleton of *Ichthyornis victor*, from Cretaceous strata, North America. (After Marsh.)

Several portions of fossil birds have been discovered in the London clay deposit of the Isle of Sheppey. One of these (which may be seen in the Natural History Museum, Table-case No. 13), the *Dasornis*, represented by a single skull, was as large as an ostrich, and probably closely related to that bird. Another, the *Argillornis*, rivalled the albatross in size. A third, the *Odontopteryx* (toothed bird), has a powerful serrated bill, well adapted for seizing its fishy prey. In the same case may be seen casts of the limb-bones of a large bird, the *Gastornis parisiensis*, from Eocene

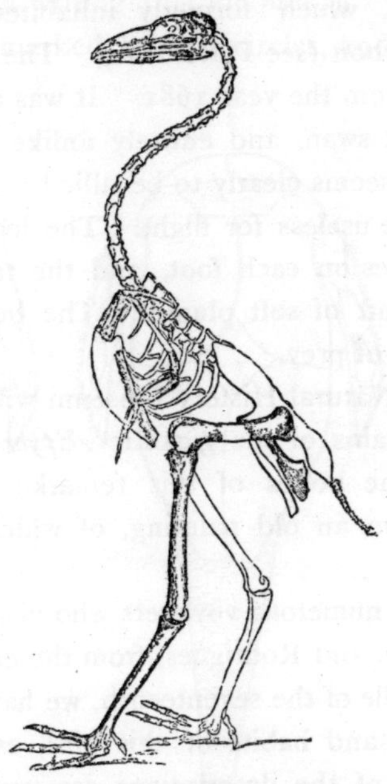

FIG. 56.—Restored skeleton of a large bird, *Gastornis Edwardsii*, from Eocene strata near Paris and Rheims. (After M. le Docteur Lemoine.) Height over seven feet. The position might be somewhat more forward.

strata near Paris; also casts of two leg-bones of another equally large bird, allied to the above, discovered in Eocene strata near Croydon, viz. *Gastornis Klaasseni*. A restoration of the French bird's skeleton is shown in Fig. 56. The genus must have been as large as an ostrich, but more robust, with some signs of affinity with geese, as well as to ratite birds such as the ostrich.

Many other orders of birds are more or less represented by fossil remains from Tertiary strata, but in most cases not so perfectly as to warrant description here. For instance, the Pliocene strata of the Sivalik hills have yielded bones of giant storks. Leaving these, we now pass on to say a few words about the Dodo, and some other birds that lived in historic times.

Every one has heard of that singular bird, now totally extinct, the *Didus ineptus*, which formerly inhabited the islands of Mauritius and Bourbon (see Plate XV.). The last record of its appearance dates from the year 1681. It was a large and heavy bird, bigger than a swan, and entirely unlike the pigeons, with which, however, it seems clearly to be allied. Its wings were so small as to be quite useless for flight. The legs were short and stout, with four toes on each foot, and the tail was extremely short, carrying a tuft of soft plumes. The beak was decidedly hooked, as in birds of prey.

Visitors to the Natural History Museum will see, in the same case with the remains of the gigantic *Æpyornis*[1] from Madagascar, some of the bones of this remarkable bird, together with a portrait from an old painting, of which we shall speak presently.

In the reports of numerous voyagers who visited the islands of Mauritius, Bourbon, and Rodrigues, from the end of the fifteenth century to the middle of the seventeenth, we have many accounts of the appearance and habits of this bird, evidently sketched from life. Some of the descriptions are very quaint; as, for example, the graphic sketch of old Sir Thomas Herbert, who saw the bird in his travels in the year 1634.

"The Dodo," he says, "comes firſt to our deſcription. Here and in Dygarrois (and nowhere elſe that I c[d] ever ſee or hear of) is generated the Dodo—(A Portuguize name it is, and has reference to her ſimplenes)—a bird which for ſhape and rareneſs might be called a Phœnix (were't in Arabia); her body

[1] See *Extinct Monsters*, p. 235 (new edit.).

is round and extreame fat, her ſlow pace begets that corpulencie; few of them weigh leſſe than fiſty pound; better to the eye than the ſtomack: greaſie appetites may perhaps commend them, but, to the delicate they are offenſive, and of no nouriſhment. Let's take her picture: her viſage darts forth melancholy, as senſible of nature's injurie in framing ſo great and maſſive a body to be directed by ſuch ſmall and complemental wings, as are unable to hoiſe her from the ground, ſerving only to prove her a bird; which otherwiſe might be doubted of: her head is variouſly dreſt, the one halfe hooded with downy blackiſh feathers; the other perfectly naked; of a whitiſh hue, as if a tranſparent lawne had covered it; her bill is very howked and bends downwards, the thrill or breathing place is in the midst of it; from which part to the end, the colour is a light greene mixed with a pale yellow; her eyes be round and ſmall, and bright as diamonds; her cloathing is of fineſt downe, ſuch as ye ſee in goſlins; her trayne is (like a China beard) of three or four ſhort feythers; her legs thick, and black, and ſtrong; her tallons or pounces ſharp; her ſtomack fiery hot, ſo as ſtones and yron are eaſily digeſted in it; in that and ſhape, not a little reſembling the Africk œſtriches: but, ſo much, as for their more certain dyfference I dare give thee (with two others) her repreſentation.[1]"

It is pretty certain that a living specimen was about the same time exhibited in England. Sir Hamon L'Estrange tells us distinctly that he saw it. His original manuscript is preserved in the British Museum, and with some blanks caused by the injury of time, of no great consequence, reads as follows:—

"About 1638, as I walked London streets, I saw the picture . . . of a strange fowl hong out upon a cloth . . . vas and myself with one or two Gen. in company went in to see it. It was kept in a chamber, and was a great fowle somewhat bigger than the largest Turky Cock, and so legged and footed, but

[1] *Travels.* 4th edit., 1677.

stouter and thicker, and of a more erect shape, coloured before like the breast of a yong Cock Tesan, and on the back of dunn or deare colour. The keeper called it a Dodo; and in the ende of a chimney in the chamber there lay an heap of large pebble stones, whereof he gave it many in our sight, some as bigg as nutmegs, and the keper told us she eats them conducing to digestion; and though I remember not how farre the keeper was questioned therein, yet I am confident that afterwards she cast them all agayne."

Probably this very specimen passed into the museum of Tradescant, who, in the *Catalogue of the Collection of the Rarities preserved at Lambeth*, dated 1656, mentions the following: "Dodar from the Island Mauritius: it is not able to flie being so bigg." Others also mentioned it.

In 1601, two fleets of Dutch ships, one commanded by Wolphart Harmansen, or Harmansz; and the other by Jacob Van Heemskerk, sailed for the East Indies, but soon separated. One of the captains, named Willen Van West-Zanen, who sailed in this fleet, and the following year sailed from Batavia with five ships, and stayed a considerable time at Mauritius. In his journal he repeatedly mentions Dodos, and his account contains some important particulars. His sailors appeared to have revelled in them, without suffering from surfeit. If his statements are correct, the bulk of the bird must have been prodigious; for he says that three or four, and in one instance two of these birds, furnished an ample meal for his men. "On the 4th of August," says his editor, "Willem's men brought fifty large birds on board the *Bruyn-Vis*, among them were twenty-four or twenty-five Dodos, so large and heavy that they could not eat any two of them for dinner, and all that remained over was salted."

The narratives of the early voyagers are, in several instances, accompanied by rude drawings; but besides these there are certain oil paintings by artists of great merit, who apparently

PLATE XV.

THE DODO (DIDUS INEPTUS).
Exterminated in the seventeenth century.

aimed only at correct representation. All, except one, closely resemble each other. Five of these are now known to exist, three of which bear the name of Roland Savery, an eminent Dutch animal-painter in the beginning of the seventeenth century; and one is by John Savery, his nephew. The best known of these pictures is the one, now in the beautiful Bird Gallery of the Natural History Museum, from which all the figures of Dodo in natural history books are taken. It was once the property of the artist George Edwards, who, in his work on *Birds*, tells us, "The original picture was drawn in Holland from the living bird, brought from St. Maurice's Island, in the East Indies, in the early times of the discovery of the Indies by way of the Cape of Good Hope. It was the property of the late Sir H. Sloane to the time of his death, and afterwards, becoming my property, I deposited it in the British Museum, as a great curiosity."

Portions of probably three distinct individuals are now extant in as many public museums; and it is remarkable, as proving the interest which the discovery of the Dodo excited in Europe, that each of these three specimens is specially referred to in the museum catalogues printed in the seventeenth century. One of these valuable fragments is the foot now preserved in the Natural History Museum. The Oxford Museum once possessed a complete stuffed specimen, bequeathed with the rest of his curiosities, by Tradescant, to Elias Ashmole, the munificent founder of the Ashmolean Museum at Oxford. Here it remained till 1755. This unhappy specimen, the last of the Dodos(!), then at least a century old, had become decayed by time and neglect, and its destruction was decreed by the trustees, who were, of course, ignorant of its priceless value. But the head and one of the feet were saved from the flames, and are still preserved in the museum. The third specimen is a skull in the Gottorf Museum at Copenhagen. Yet it cannot be doubted that if careful search were made in the caves and superficial deposits of Mauritius, many more bones might be discovered. Entire skeletons of the Dodo and the Solitaire are exhibited in the Bird Gallery of the Natural History Museum, among the Pigeons.

There is no reason why a pigeon should not become so modified, as the result of a change in external circumstances, as to be incapable of flight. Now, we are told that Mauritius, an island forty miles in length, and about a hundred miles from the nearest land, was, when discovered, clothed with dense forests of palms and various other trees. A bird adapted to feed on the fruits produced by these forests would, in that equable climate, have no occasion to migrate to distant lands; it might revel in the perpetual luxuriance of tropical vegetation and could have but little need of locomotion. Why, then, should it trouble to use its wings? Such a bird might wander from tree to tree, tearing with its powerful beak the fruits which strewed the ground, and digesting their stony kernels with its powerful gizzard, enjoying tranquillity and abundance, until the arrival of man destroyed the balance of animal life, and put an end to its existence.

In a second island forming one of the Mascarene group, viz. that known as Rodriguez, situated about three hundred miles east of Mauritius, early explorers found another bird living. This was the Solitaire (*Pezophaps solitaria*), a near relative of the Dodo. It differed from the latter in having longer legs and neck, and its bill not so much curved. The last recorded appearance was in 1693. Rodriguez seems to have remained in a desert and uninhabited condition until 1691, when a party of French Protestant refugees settled upon the island, and remained there for two years. Their commander, François Leguat, a man of intelligence and education, has left a highly interesting account of their adventures, and of the various productions of the island.[1]

"Of all the Birds in the Island," he says, "the most remarkable is that which goes by the name of the *Solitary*, because it is very

[1] *A New Voyage to the East Indies by Francis Leguat and his Companions.* 12mo. London, 1708.

seldom seen in Company, tho' there are abundance of them. The Feathers of the Males are of a brown grey colour, the Feet and Beak are like a Turkey's, but a little more crooked. They have scarce any Tail. . . . Their Neck is straight, and a little longer in proportion than a Turkey's when it lifts up its head. . . . They never fly, their Wings are too little to support the weight of their bodies; they only serve to beat themselves, and flutter when they call one another. . . . From *March* to *September* they are extremely fat, and taste admirably well, especially while they are young: some of the Males weigh forty-five pounds."

The females he describes as being wonderfully beautiful, and walking with much stateliness and grace: after which he proceeds as follows: "Tho' these Birds will sometimes very familiarly come up near enough to one, when we do not run after them, yet they will never grow tame. As soon as they are caught they shed Tears without Crying, and refuse all manner of Sustenance till they die."

The figure given by Leguat shows that it was a very different bird from the Dodo. Its nest was a heap of palm-leaves, and it laid but one egg, three times a year.

In conclusion, we will quote Leguat once more: "Our Wood-Hens," he says, "are fat all the year round, and of a most delicate taste. Their colour is always of a bright grey, and there's very little difference in the plumage between the two Sexes. They hide their nests so well that we cou'd not find 'em out, and consequently did not taste their eggs. They have a red list about their eyes, their beaks are straight and pointed, near two inches long, and red also. They cannot fly, their fat makes 'em too heavy for it. If you offer them anything that's red, they are so angry that they will fly at you to catch it out of your hand, and in the heat of the combat we had an opportunity to take them with ease."

The volcanic island of Bourbon, lying about a hundred miles south-west of Mauritius, was certainly inhabited by two species of birds, whose inability to fly, and consequent rapid extinction, brings them into the same category with the Dodo of Mauritius and the Solitaire of Rodriguez. Bourbon was discovered between the years 1502 and 1545, by Mascaregnas a Portuguese, who called the island by his own name, but seems to have left no other record of his visit. In the seventeenth century, travellers such as Bontekoe and Carré speak of a big short-winged bird which they saw here, and which from their descriptions appears to be the Solitaire, for they said its colour was white. But in 1669, the year after Carré's visit, a French colony was sent from Madagascar to Bourbon under M. de la Haye. One of the party, who calls himself the Sieur D.B., has left an interesting account of the expedition. He distinctly alludes to a second species of short-winged bird, in the following words :

Oiseaux bleus, the size of *Solitaires*, have the plumage wholly blue, the beak and feet red, resembling the feet of a hen. They do not fly, but they run extremely fast, so that a dog can hardly overtake them ; they are very good eating." We must suppose it escaped the notice of the earlier voyagers.

On reviewing the various historical and other evidences, which have been carefully brought together by Mr. H. E. Strickland, it seems clear that the three oceanic islands, Mauritius, Rodriguez, and Bourbon, which, though somewhat remote from each other, may be considered as forming one geographical group, were inhabited, until the time of their colonization, by three, or perhaps four, distinct but allied flightless birds. These constitute a distinct family, the *Dididæ*, of naturalists, allied to the pigeons, but very isolated. Before men came thither they flourished abundantly; but when cats, dogs, and pigs were introduced by human agency, and man also himself sought them for food, their fate was sealed, and they rapidly became exterminated. The fact that these perfectly defenceless creatures survived in great abundance to a recent period in these three islands only, while as far as we know they never existed in any other countries whatever, certainly confirms other evidence which might be brought forward, did space permit, to show that they are very ancient but truly "oceanic" islands—*i.e.* they never formed part of any continent and subsequently became separated, as Madagascar did from Africa, or the British Isles from Europe. From what is already known of bird life in the Miocene period, it seems probable that the origin of this peculiar group of the Dodo and its allies dates back to early Tertiary times.

"If we suppose," says Mr. Alfred Russell Wallace,[1] "some ancestral ground-feeding pigeon of large size to have reached the group by means of intervening islands afterwards submerged, and to have henceforth remained to increase and multiply unchecked by the attacks of any more powerful animals, we can well understand that the wings, being useless, would in time become almost aborted. It is also not improbable that this process would be aided by natural selection, because the use of wings might be absolutely prejudicial to the birds in their new home. Those that flew up into trees to roost, or tried to cross over the mouths of rivers, might be blown out to sea and destroyed, especially during the hurricanes which have probably always more or less devastated the islands; while, on the other hand, the more bulky and short-winged, who took to sleeping on the ground in the forest, would be preserved from such dangers, and perhaps also from the attacks of birds of prey, which may always have visited the island."

Whether this is a complete account of how the change took place or not, we may take it for certain that their abnormal development could not have taken place except under complete isolation and freedom from the attacks of enemies. In other countries, like Europe, Africa, or Asia (but not in Australia or New Zealand), there was, during Tertiary times, an abundance of carnivora to keep down the rest of the animal creation, and prevent any ambitious attempts on the part of pigeons

[1] Alfred Russell Wallace, *Island Life*, p. 407. (1880.)

to increase in size; so that Dodoism, if we may be allowed to use the expression, would be impossible.

The Samoa Isles in the Pacific recently possessed a large and handsome kind of pigeon, of richly coloured plumage, which the natives called *Manu-mea*, but to which modern naturalists have given the name of *Didunculus strigirostris.* It was, both by structure and habit, essentially a ground pigeon, not, however, so exclusively but that it fed and roosted too, according to Lieutenant Walpole, among the branches of tall trees. Mr. T. Peale, the naturalist of the United States Exploring Expedition, who first described it, informs us that, according to the tradition of the natives, it once abounded; but after the introduction of cats, dogs, and rats, this bird was soon quite exterminated. So rare had the bird become during the stay of the expedition, that only three specimens could be procured, and, of these, two were lost by shipwreck.

When Norfolk Island, in the Southern Ocean, was first discovered, its tall forests were inhabited by a remarkable parrot with a very long and slender hooked beak, which lived upon the honey of flowers. Mr. Gould, the ornithologist, on visiting Australia, found this bird—known as *Nestor productus*—entirely limited to Philip Island, a little island not far off. The war of extermination had been so successfully carried on through the agency of man that, in the larger island, only a few specimens in cages were to be seen.

Mr. J. H. Gurney has given us a description of it. He says:[1] "I have seen the man who exterminated the *Nestor productus* from Philip Island, he having shot the last of that species left on the island. He informs me that they rarely made use of their wings, except when closely pressed; their mode of progression was by the upper mandible; and whenever he used to go to the island to shoot, he would invariably find them on the ground, except one, which used to be sentry on one of the

[1] *Zoologist*, p. 4298.

lower branches of the *Araucaria excelsa*; and the instant any person landed, they would run to those trees and haul themselves up by the bill, and, as a matter of course, they would there remain till they were shot, or the intruder had left the island. He likewise informed me that there was a large species of hawk that used to commit great havoc amongst them, but what species it was he could not tell me."

That fine bird the Great Auk is believed now to be quite extinct. But it was once abundant in Northern Europe. The natives in the Orkney Island informed Mr. Bullock, on his tour through these islands some years ago, that only one male had been seen for a long time since, which had regularly visited Papa Westra for several seasons. The female, which the natives call the queen of the Auks, was killed just before Mr. Bullock's arrival. The king, or male, was chased by him for several hours, but without success, so rapid was its course under water. It was afterwards caught and sent to him, and is now in the national collection. In size it was rather larger than a goose. Mr. Bullock's specimen was taken in 1812; another was captured at St. Kilda in 1822; another was picked up dead near Lundy Island in 1829; and yet another was taken in 1834, off the coast of Waterford. On the north coast of Europe the bird was equally rare, not more than two or three having been procured during the present century. Two hundred years ago it was by no means rare on the shores of New England; and off the great fishing-banks of Newfoundland it was very abundant. "During the sixteenth and seventeenth centuries," says Dr. Charlton, "these waters, as well as the Iceland and Faroe coasts, were annually visited by hundreds of ships from England, France, Spain, Holland, and Portugal; and these ships were actually accustomed to provision themselves with the bodies and eggs of these birds, which they found breeding in myriads on the low islands off the coast of Newfoundland. . . . It was only necessary to go on shore armed with sticks, to kill as many as they choose. The birds were so stupid that they allowed themselves to be taken up, on their own proper element, by boats under sail. . . .

"In 1841, a distinguished Norwegian naturalist (too early, alas! lost to science), Peter Stuwitz, visited Funk Island, or Penguin Island, lying to the east of Newfoundland. Here, on the north-west shore of the island, he found enormous heaps of bones and skeletons of the Great Auk, lying either in exposed masses, or slightly covered by the earth. On this side of the island, the rocks slope gradually down to the shore, and here were still standing the stone fences and enclosures into which the birds were driven for slaughter."[1]

It is often remarked that men and women who are specially gifted in certain directions are found to be wanting in some other ways; for there is compensation in all things, and no one individual can excel in all the gifts and graces at once. In order to achieve great results in one field, we must leave others untouched. The artist or musician is not expected to excel in physical strength; nor is the athlete expected to show any great gift for painting or for music. Each must be "specialised" in his own way, and should be content if he can bring to perfection his one gift. So it is in the animal kingdom; some are gifted in one direction, some in another. Fishes excel in swimming, mammals in walking or running, and birds in flying. Each has conquered in his own realm. Now the bird class has conquered the air; but, in order to do this, and to fly from place to place with ease and rapidity, they have had to make no small sacrifice; for they have turned their fore limbs, which the reptile uses for crawling, into wings for flying. It is true they have much to compensate them for this sacrifice in the vigour and happiness of their lives; in the marvellous gift of song, and in the social qualities they display—to say nothing of their wisdom, which is proverbial. Still the greatest triumphs are

[1] *Transactions of the Tyneside Natural History Society.*

reserved for those creatures which have remained on the ground, and continued to use their fore limbs in the ordinary way. Miss Buckley (Mrs. Fisher) has, in one of her best books,[1] expressed this idea very happily. We will therefore quote the passage, as we find ourselves unable to put it in better words: "Thus the birds," she says, "with their feathery covering and powerful wings, have left their early friends, the reptiles, far behind. Taught by their many dangers, many experiences, and many joys, they have become warm-hearted, quick-witted, timid or bold, ferocious or cunning, deliberate as the rook or passionate as the falcon, according to the life they have led; or, in the sweet, tender emotions of the little song-birds, have learned to fill the world with love and brightness and song. . . .

"Yet we cannot but feel that, happy as a bird's life may be, it still leaves something to be desired; and that, with their small brain, and their front limbs entirely employed in flying, they cannot make the highest use of the world. The air they have conquered; and among the woods and forests, over the wide sea, and above the lofty mountains, they lead a busy and happy existence, bringing flying creatures to their highest development, and showing how life has left no space unfilled with her children. Yet, after all, it is upon the ground, where difficulties are many, conditions varied, and where there is so much to call for contrivance, adaptation, and intelligence, that we must look for the highest types of life; and while we leave the joyous birds with regret, we must go back to the lower forms among the four-footed animals, in order to travel along the lines of those that have conquered the earth and prepared the way for man himself."

[1] *Winners in Life's Race*, p. 179.

CHAPTER IX.

TAPIR-LIKE ANIMALS AND ELEPHANTINE MONSTERS.

"To lead the mind of man towards its noble destination—a knowledge of truth; to spread sound and wholesome ideas among the lower classes of the people; to draw human beings from the empire of prejudices and passions; to make reason the sole arbitrator and supreme guide of public opinion."—CUVIER, on the object of Science, in his *Report to the Emperor Napoleon on the Progress of the Natural Sciences since the Year* 1789.

IT has often been said that the Primary era was an "Age of Fishes," the Secondary era an "Age of Reptiles," and the Tertiary an "Age of Mammals." There is, however, a danger lest beginners should be deceived by broad statements of this kind, which must not be pressed too far. All that is meant in this case is, that, first of all, fishes were the dominant type; then reptiles, and then mammals.

We have shown in earlier chapters of this book, as well as in our previous work, that fishes abounded in the waters of the Primary or Palæozoic era, while air-breathing amphibians appeared towards its close; that reptiles flourished vigorously all through Secondary or Mesozoic times: and now we are about to show that a marvellous outburst of mammalian life appears to have taken place very early in the Tertiary or Cainozoic era. All this is in accordance with the "Law of Progress" throughout past times, and strongly confirms the theory of Evolution.

The student of geological history soon discovers that mammals of some kind, or kinds, *did* exist throughout the Secondary era—

although, judging from their imperfect remains, they must have been of a low type.[1] But the great advance in our knowledge of the world's extinct races which has taken place during the last quarter of a century ought to make us careful with regard to broad statements of the above kind, which may require to be modified by future discoveries.

Now, the tendency of all the later results of Palæontology is to show that some of the higher types of life appeared on the earth a good deal earlier than was formerly supposed. Thus the discovery of the jaw-bones of small mammals (possibly marsupials) in the Purbeck strata, in 1854, and afterwards in the Stonesfield slate (both of Jurassic age), came as a surprise to most geologists. Again, it is not long since birds were believed to have come into existence only in Tertiary times: *now* we have the Jurassic bird *Archæopteryx* (see p. 152), and Professor Huxley has recorded his opinion that some of us may live to see a fossil mammal of the remote Silurian age! It is, therefore, only prudent for geologists to be on their guard against assuming that *all* the Secondary mammals were of low types and few in numbers, as might be inferred from the phrase "Age of Mammals." That mammals, however, developed vigorously in many directions during the Tertiary era cannot be doubted. So far, then, the expression conveys important truth; but in science, as in other things, the "unexpected" often happens, so it is better to be prepared for surprises.

And here an interesting question naturally suggests itself, viz. What cause, or causes, operated to bring about such rapid and vigorous expansion of certain classes, such as the reptile and the mammal? Now, although this question cannot be fully answered, there are nevertheless certain highly suggestive known facts which seem partly to solve the mystery. We may suppose that, for long

[1] Perhaps some of the jaw-bones from Triassic and Jurassic strata represent creatures as low down in the scale as the Monotremes, represented at the present day by the Duck-bill *Platypus* and the *Echidna* of Australia.

ages before the Tertiary era, mammals had remained in a stationary and sluggish condition; then some change, or changes, took place, which, as it were, gave them a fresh start. What was the nature of such change? Let us look at the living world, for a moment, as it is now; for in Geology the past must be interpreted by the present. Naturalists know well that any change in climate (*i.e.* in temperature or atmospheric conditions), or in the configuration of land and sea, is sure to effect the balance, or equilibrium, of life. Some creatures will gain thereby, others will lose, and may in consequence become extinct. Here we have evidently a potent cause at work, which may be summed up under the phrase *geographical changes*. That such have been constantly occurring in the past is beyond doubt, and it is equally certain that they have brought about vast changes in the organic world; for plants and animals are highly sensitive to the conditions around them. Take away the Gulf Stream from the North Atlantic, and in time many highly important changes in the fauna of that sea would inevitably follow; flood the desert of Sahara, or a part of it, and you would bring life and fertility to the surrounding districts, making the very desert to blossom as the rose; remove a natural barrier, such as a channel or sea between two masses of land, or a mountain chain separating two countries, and you would at once start a series of migrations and other changes the consequences of which would be almost endless; cut down a forest, and you diminish rainfall, thereby altering the whole climate of the region. It is not surprising to learn, in view of these facts, that the periods of greatest geological disturbance—affecting the distribution of land and water, and perhaps lifting up whole mountain ranges—have always been followed by great changes in the organic world, and the rapid evolution of new types of life.

The geologist can easily call to mind examples of this principle. Thus, the Coal-measure period was followed by great geographical changes, and, consequently, we find very different faunas in the rocks subsequently deposited together, with the coming in of higher types. Again, at the close of the Cretaceous period such changes took place on a large scale, sea becoming dry land and the land becoming sea, and consequently we are at once introduced, in the Tertiary deposits, to an amazing number of newer and higher types, which had, as it were, been waiting for a chance to "improve themselves," and rise in the scale of being.

It would be easy to illustrate this truth from human affairs; for instance, in the history of the growth of the Anglo-Saxon race in America and the colonies, where the old stock has flourished vigorously in new fields. Whether they will develop a "higher type" remains to be seen; but in America, we see the production of a somewhat novel one. The discovery of the New World and the building of huge ships, making, as it were, a ferry across the Atlantic, has led to "migration" on a large scale. Nature, however, brings about migrations by changing the geography of the globe. For example, Australia was once united to Asia, and creatures living ages ago in Asia found their way into Australia; them came elevation, and consequent separation. Europe and America were once united towards the North; Africa and India formerly had a belt of land connecting them, as proved by the Gondwana beds and their fossils. Now, when by any such means a group of animals migrates into a new territory and finds there a fresh field, with less powerful competitors and an abundance of food, it is obvious that they will enjoy a "good time;" the result of which seems to be that they flourish vigorously and expand in many directions, thus producing new branches of the Tree of Life. We all know how the rabbits took to Australia and throve only too vigorously; and we have been informed that already they have begun to alter their habits and to climb trees, for which purpose they are developing longer claws! Perhaps we shall in time see a new race of these little rodents "evolved" in that country.

There is, however, another cause which may be of even greater importance, although it cannot be so easily understood,—and that is, *internal changes in the animal itself.* All the creatures we see around us are constantly varying, and have done so in the past—now in one direction, now in another. The cause of "variation" is one of the unsolved problems of modern Biology, or the study of life; but we do know that a variation occasionally happens to be of such a kind as to make a radical change in the organism, and to fit it for new conditions of life better than its comrades of the same species. As an example of such a change, whether acquired at once by some individual, or only slowly brought about after many generations, we may take the case of those mammals of the higher orders, which have the power of retaining their young for a longer time within their bodies than either the monotremes or the marsupials.[1] This character, together with others that necessarily go with it, made the creature thus favoured more powerful, *i.e.* more safe and better able to take care of itself and its young ones, so that it had a better chance in the "struggle for existence." Certain it is that those mammals which have this habit have triumphed over the marsupials and reptiles. Hence it came about that, as soon as this important variation occurred in the main stock, an immediate impulse was given to its development. They became more active, more numerous, and more powerful. This was probably one of the ways in which mammals ascended in the scale of life and branched out into various orders, such as insectivora, ungulates, rodents, carnivora, etc.

On the other hand, the same reasoning would lead us to infer that the reptile class flourished well during the Mesozoic era because they were the first air-breathing vertebrates (besides having this new habit of breathing air directly instead of from water as the fishes do) and therefore had the land all to them-

[1] So called because they carry their young ones in a pouch (marsupium); for example, the Kangaroo.

selves. Hence it is not surprising to find that throughout that long era they developed many branches, the majority of which, however, are now extinct.

Similarly, the class of birds probably underwent great expansion at the same time, because they were conquering a new province, the great ocean of air, and, in order to do this, had acquired new structures and new habits, giving them advantages over their old ancestors the reptiles.

Numerous orders of insects appear in the Jurassic and Cretaceous deposits; and it has been suggested that their evolution may have been due to the appearance, for the first time in the world's history, of flowers on the land, and the habit of certain insects of feeding from them.

Lastly, some geologists have lately attributed the great expansion of mammalian life in Tertiary times to the probable fact that grassy pastures only began to grow on the earth in the early days of that era. A few students of the evolution of plant life have lately arrived at the conclusion that the grasses (*gramineæ*) did not exist in the previous geological periods. Here, then, we seem to have a powerful cause suggested; and, if it is a true one, "pastures new," in a literal sense, were opened out for the budding mammalian line to feed on. What endless possibilities were thus within their reach! Here was better and more nourishing food for them than the ferns, reeds, or cycads, to which they had previously been limited. So far, these suggestions help us better to understand how it was that the mammals in Tertiary days struck out so many new lines; but we must bear in mind that our knowledge of these mysterious operations is at present very limited, and doubtless new causes will be discovered as time goes on.

It would be impossible within the limits of a single chapter to give an account of all the leading orders of Tertiary mammals. We must therefore limit our present remarks to two or three, reserving others for future chapters.

Beginning with certain Tapir-like animals that flourished during Eocene times, we will then pass on to consider some curious creatures of larger size, some of which are true elephants, while others seem to be related to that order (the proboscidians) as well as to the rhinoceros.

The discovery, in the early part of the present century, of the rich treasures imbedded in the Tertiary strata of the Paris basin, and the consummate skill with which they were interpreted and restored by the immortal Cuvier, gave a very great impulse to the study of Geology. His restorations became patterns for others, like Owen, Huxley, Marsh, Cope, Gaudy, and many more, who have worked on the lines he laid down. The mammalian remains brought to Cuvier by numerous collectors were very imperfect and fragmentary—detached bones and teeth, with occasionally some portion of a skeleton. The success with which he put them into order, and built up therefrom the long-lost types of Eocene days, was due largely to his wonderful knowledge of living animals, but partly also to his *Law of Correlation.*

One of Cuvier's triumphs was the restoration of the *Palæotherium*,[1] a tapir-like animal, from fragmentary remains found in the strata of the Paris basin, chiefly at Montmartre; his conclusions being afterwards verified by the discovery of a nearly complete skeleton (see Fig. 57). The molar teeth of this animal somewhat resembled those of the rhinoceros. The skull shows that it had a short proboscis. The toes, of which there were three on each foot, ended in small hoofs, the middle toes being the largest. Its eye was small, and the head rather large. Several species have been determined, varying from the size of a sheep (*P. curtum*) to that of a horse (*P. magnum*), and it had much in common with both the horse and the rhinoceros. *P. medium* was rather smaller than an American Tapir: *P. minus* was a small and elegant species, of which the fresh-water Eocene beds of the Isle of Wight have yielded remains.

[1] Greek—*palaios*, ancient; *therion*, wild beast.

TAPIR-LIKE ANIMALS. EOCENE PERIOD.

Xiphodon. *Anoplotherium.* *Palæotherium.*

PLATE XVI.

A complete specimen, discovered in 1874, with the outline of the body indicated in the rock, was of a slenderer build, and with a longer neck than in Cuvier's restoration, on which our artist's drawing in Plate XVI. is based; but the earlier discoveries do not agree with this, and the reader can easily see for himself that the skeleton shown in Fig. 57 could not have had a longer neck without at the same having longer legs, and probably thinner bones. We have therefore adhered to the now familiar outline of the Palæothere as restored by Cuvier and since copied into almost every book on Geology. Doubtless there were many species

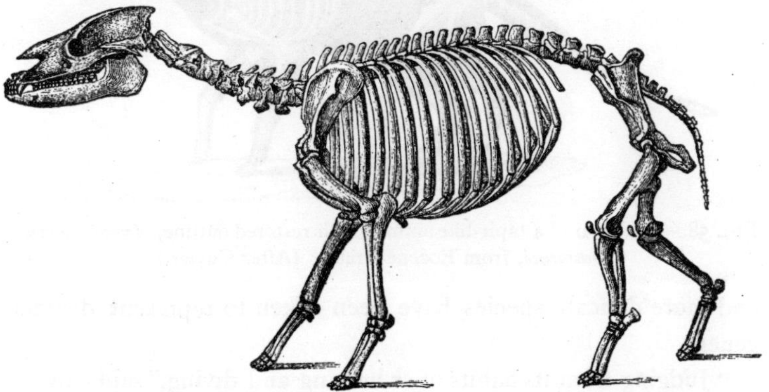

FIG. 57.—Skeleton of a tapir-like animal, *Palæotherium magnum*, from Eocene strata, near Paris. (After Gaudry.)

existing at the time, and this later discovery probably represents one of the slenderer sort.

Such then, in all probability, was the Palæothere, a creature which lived in herds in the valleys of the plateau surrounding the ancient lake-basins of Orleans and Argenton; in the department of Gironde; in the Isle of Wight; and in various parts of Europe.

Its contemporary, the *Anoplotherium*,[1] so called from its apparently defenceless state, is represented in the same plate. This animal was of a lighter and more elegant form, and its limbs

[1] Greek—*alpha*, privative; *hopla*, arms; *therion*, beast.

ended each in two digits, only terminating in hoofs. As in the Palæothere, the jaws contained forty-four teeth, but there was no interval in the series. There are suggestions in its framework of the modern ruminant type;[1] thus it resembled the ox in having an equally divided hoof, or rather two hoofs. Fig. 58, in which the skeleton is seen, shows that it possessed a long tail. In size it was about equal to a fallow deer, and the long tail has been supposed, perhaps erroneously, to indicate aquatic habits. Small

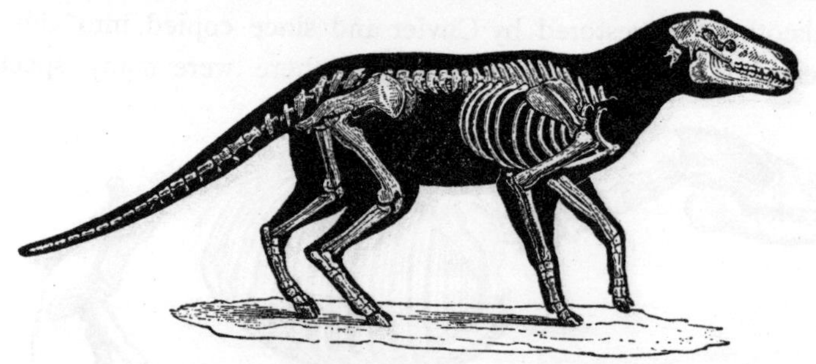

FIG. 58.—Skeleton of a tapir-like animal, with restored outline, *Anoplotherium commune*, from Eocene strata. (After Cuvier.)

and more delicate species have been taken to represent distinct genera.

"Judging from its habits of swimming and diving," said Cuvier, "the *Anoplotherium* would have the hair smooth like the otter; perhaps its skin was even half naked. It is not likely, either, that it had long ears, which would be inconvenient in its aquatic kind of life; and I am inclined to think that, in this respect, it resembled

[1] Cuvier divided all the hoofed animals (ungulates) into two orders, pachyderms and ruminants. The former is a heterogeneous order, and has since been abandoned; but the ruminants have been regarded as one of the most distinct of mammalian orders, for they are separated from all other animals by having horns and hoofs in pairs, the absence of upper front teeth, complex stomachs, and the habit of ruminating or "chewing the cud." Professor Owen showed that ungulates should be classified by the structure of their feet. He therefore divided them into odd-toed (perissodactylate) and even-toed (artiodactylate), and he placed elephants in a separate order—the Proboscidia.

the hippopotamus and other quadrupeds which frequent the water much." But was it really aquatic?

Xiphodon,[1] so named by Cuvier on account of the shape of its teeth, was a small and delicate animal, long, with slenderer limbs than the *Anoplotherium;* its feet were provided with two toes, and the tail was short (see restoration in Plate XVI.). *X. gracilis* was obtained from the lignites of Débruge, near Apt. It was some three feet high, and of about the size of a chamois, but lighter in form, and with a smaller head. Cuvier says of this creature, "Its course was not embarrassed by a long tail; but, like all active herbivorous animals, it was probably timid, and with large and very mobile ears, like those of the stag, announcing the slightest approach of danger. Neither is there any doubt that its body was covered with smooth hair, and consequently we only require to know its colour in order to paint it as it formerly existed in this country, where it has been dug up after so many ages." Instead of resorting to rivers and lakes, this graceful little creature probably kept to the dry land and fed upon aromatic herbs.

Discoveries, chiefly made since Cuvier's day, have shown that these ancient herbivorous mammals had carnivorous enemies, which doubtless kept down their numbers—not lions and tigers, but certain lower and less "specialised" creatures, from some of which the latter are descended.

Chœropotamus,[2] the Water-hog, was another contemporary, having a good deal of analogy with the living Peccari, though much larger; its feet have not yet been found, but they probably had four toes.

Anthracotherium[3] (see Fig. 59), so named from having been found in a bed of anthracite, or lignite, near Savone in France, doubtless dwelt in swamps and marshes, and belongs to the same

[1] Greek—*xiphos*, sword; *odous*, *odontos*, tooth.
[2] Greek—*choiros*, a young pig; *potamos*, river.
[3] Greek—*anthrax*, coal; *therion*, beast.

family as *Chæropotamus*. It appears to have resembled both a pig and a hippopotamus (see restoration, Plate XVI.).

In the strata of the Paris basin there are three masses of gypsum separated by intervening deposits of fine marl, which breaks up into fine layers, or laminæ. In the top deposit, in the valley of Montmorency, M. Desnoyers discovered, in 1859, many foot-prints of animals, occurring at no less than six different levels. The gypsum to which these marls belong varies from thirty to fifty feet in thickness. Sir Charles Lyell visited the quarries soon after the discovery was made known, with M. Desnoyers, who showed him large slabs in the museum at Paris, where, on the

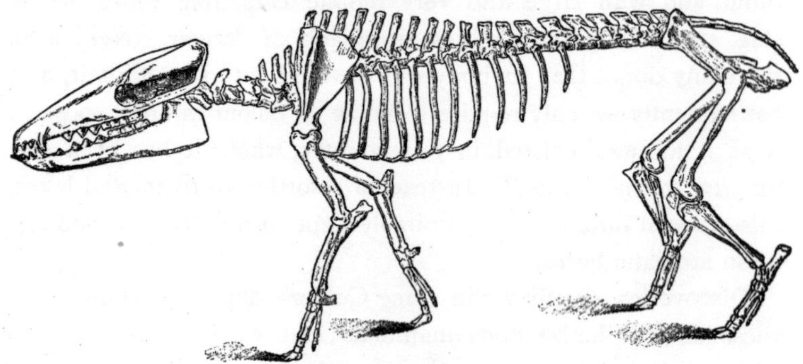

FIG. 59.—Skeleton of *Anthracotherium*, from Lower Miocene strata. (After Kowalevsky.)

upper planes of stratification, the indented footmarks were seen, while corresponding casts in relief appeared on the lower surfaces of the gypsum beds immediately above. Each of these thin films of marl, before being hardened by pressure of overlying rocks, was in the state of mud; on this mud the animals of this Eocene period once walked or "made tracks," as Americans say, and the impressions penetrated to the gypsum below, which must then have been in a soft condition. Here, then, we have an old haunt of early Tertiary forms of life; and this haunt, the geological record tells us, was probably near a lake, or several small lakes,

TAPIR-LIKE ANIMALS. EOCENE PERIOD.

Palæosyops. *Anthracotherium.*

PLATE XVII.

communicating with each other, on the shores of which numerous herbivorous creatures wandered, together with beasts of prey which occasionally devoured them. This is clearly proved by the tooth-marks detected by palæontologists on the bones and skulls of Palæotheres entombed in the gypsum.

The nearest living ally of our Palæothere, that once inhabited Europe in great numbers, must be sought for a long way from home; and it is highly instructive to observe that the further we wander away from the present down "the corridors of time," the further we must travel, geographically, to find creatures at all matching those that lived in early ages of the world's history. Thus, the Tapir from Sumatra, or Central and South America, is the nearest living relative of the *Palæothere;* while, if we go back still further—into Jurassic times,—we must fetch marsupials from Australia to match some of the little mammals whose jawbones are found in some of the deposits of that age. To the geologist and naturalist, South America and Australia are countries which have, as it were, lagged behind; for their faunas are not "up to date," as the saying is; and thereby hangs a long story—into which we must not be led now,—and one which shows how closely the sciences of Geology, Natural History, and Geography are interwoven.

As already stated, Cuvier's great results in restoring Eocene animals were due, in a large measure, to the use he made of his principle of "correlation;" of which he writes: "I doubt whether I should ever have divined, if observation had not taught me, that the ruminant hoofed beasts should all have the cloven foot, and be the only beasts with horns on the frontal bone."[1] Again, it is found—though no one knows why—that only those hoofed animals which have their hoofs in one or two pairs, have horns in one or two pairs on the frontal bones: whilst those with three hoofs, if they have horns at all, have either one or two placed one behind the other (example, the Indian rhinoceros with one

[1] *Ossemens Fossiles*, tom. i. p. 184. (1834.)

horn and the African two-horned rhinoceros). There must be secret reasons for these curious facts—who shall win immortality by discovering them? Other relationships, such as that between teeth and hoofs, might be given; but the above example is sufficient for our present purpose. In dealing with this subject, however, it should be mentioned that Cuvier's law is not infallible, and does not always apply to some of the ancient and generalised types discovered since his day.

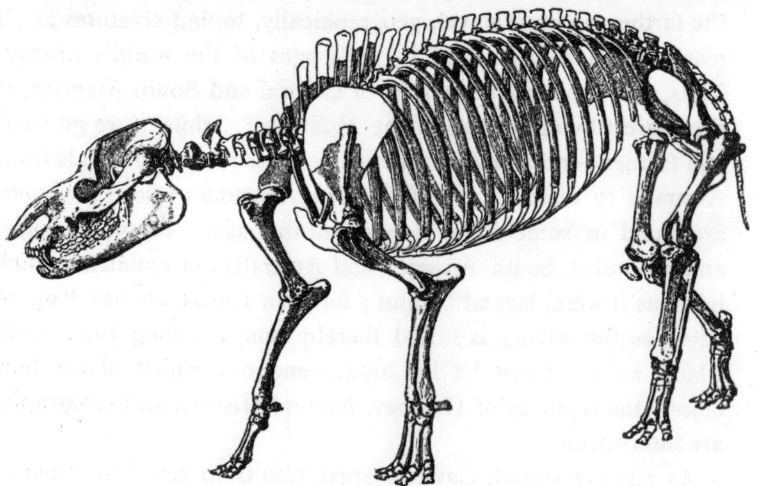

FIG. 60.—Skeleton of a tapir-like animal, *Palæosyops*. (Restored, after C. Earle.)

Our account of Eocene Tapir-like animals would not be complete without some mention of an interesting animal allied to the Palæothere, discovered in Tertiary strata in North America, viz. the *Palæosyops*,[1] so named from its pig-like face. In general features it strongly resembled the Tapir, had a stout body, with slender tail, and short neck, compensated by a proboscis of considerable length.

Our restoration, seen in Plate XVII., is based upon a complete

[1] Greek—*palaios*, ancient; *sus*, pig; *ops*, face.

restoration of the skeleton, lately published in America, by Mr. C. Earle[1] (see Fig. 60). That drawing represents a great amount of labour and is the result of much collecting on the part of several workers for a good many years. It is founded on specimens, by no means complete, in the museum of the Academy of Natural Sciences of Philadelphia; in the museum of Princeton College; in the collection of Professor Cope; and some of Professor Marsh's specimens in the museum of Yale College. The name of the late Dr. Leidy, a pioneer in American palæontology, is associated with this genus, and the collection in the museum of the Academy of Natural Science, Philadelphia, contains many of the original specimens from which Leidy first gave to the scientific world the knowledge of the existence of this creature. Since he, in 1870, described the genus from a few fragments of teeth found in Wyoming, a great advance has been made in the knowledge of its anatomy. Of late years Professors Scott and Osborne have made large collections of its bones in their well-known western exploring expeditions. *P. borealis*[2] comes from the Wind River beds, and *P. major* from the Bridger beds of the American Eocene.[3]

We pass on now to give a brief account of a strange elephantine creature that lived in Eocene times, both in America and Europe, the *Coryphodon*[4] (so named from its teeth), a restoration of which is seen in Plate XVIII. The complete skeleton, as restored quite recently by Professor Marsh, is shown in Fig. 61.[5] The history

[1] *Journ. Ac. Sciences, Philadelphia*, 2nd series, vol. ix., part iii., p. 314.

[2] To save space we have allowed the *Palæotherium* and *Anthracotherium* to appear together on the same plate, although one is an American form, the other European.

[3] The divisions of the American Eocene strata are as follows, in descending order:—

Uinta.
Bridger.
Wasatch.
Puerco.

[4] Greek—*korufé*, a ridge; *odous*, *odontos*, tooth. *Vide* p. 189.

[5] From Professor Marsh's paper, *Amer. Journ. Science*, xlvi. (1893), p. 325, a copy of which he kindly sent to the author.

of this remarkable animal, so long shrouded in obscurity, is worth recording here. The specimen on which the genus was founded by Sir R. Owen, in 1846, is unique, and was dredged up from the bottom of the sea, between St. Osyth and Harwich, on the Essex coast. It appears to have been washed out of the London Clay formation, as many other fossils have been; and consists of the right branch of the lower jaw. It may interest the reader to know that this distinguished naturalist confessed that he had seldom felt more misgiving in regard to a conclusion based on a single tooth or bone than that which he arrived at after a careful study of this specimen. Its smaller and less obvious features

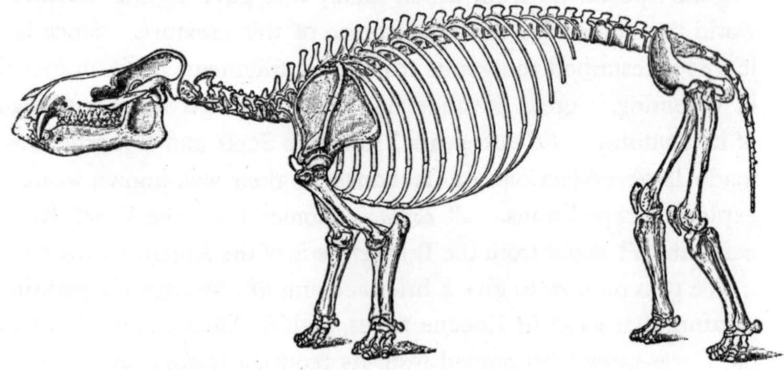

FIG. 61.—*Coryphodon hamatus.* (Restored, after Marsh.) Length about six feet.

carried conviction to him against the showing of the larger and more catching ones. But although some naturalists for a time thought he had mistaken the fore for the back part of the jaw, yet his conclusion proved to be correct. His experience taught him that the less obvious points, which require searching out, frequently, when their full meaning has been grasped, guide to a right interpretation of the whole. "It is as if truth were whispered," he says, "rather than outspoken by Nature."

The first additional evidence which Sir R. Owen obtained of the true nature of this ancient mammal was furnished by a fossil

AN ANCIENT MAMMAL (CORYPHODON HAMATUS). EOCENE PERIOD.
Length 6 feet.

Plate XVIII.

canine tooth brought up from a depth of a hundred and sixty feet out of the Plastic Clay during the operation of sinking a well in the neighbourhood of Camberwell, near London. This circumstance caused Sir R. Owen to remind his readers of the old proverb, "Truth lies at the bottom of a well." It was nearly three inches long, and evidently belonged to a large hoofed mammal. With regard to the teeth, he remarks that their broad-ridged and pointed grinding surfaces indicate that they were intended to be applied to the coarser kinds of vegetable substances.

According to Owen, certain fossils from the lignite deposits of Soisson, Laon, and Meudon, in France, belong to *Coryphodon*. Speaking of a tooth from Soisson, Cuvier said that the entire skeleton was found indicating an animal almost as large as a bull, but that the workmen employed in the sand-pit (*sablionere*) unfortunately preserved only that one tooth. The first specimen of *Coryphodon* discovered in America was found in 1871, near Evanston, Wyoming, by Mr. William Cleburne, while engaged as surveyor for the Union Pacific Railroad, who secured chiefly its teeth and vertebræ. More or less perfect specimens were afterwards obtained by Cope during his explorations in New Mexico, under the survey of Captain Wheeler. Regarded merely as a fossil, it is characteristic of the Lower Eocene of Europe and America; in North America it is confined to the Wasatch and Wind River epochs, and is absent from the Upper Eocene of both countries.

Coryphodon is particularly interesting on account of the primitive features of its skeleton; the brain was very small and of a low type. Professor Cope places it, with its allied forms, in a separate order, to which he gives the name Amblypoda, on account of their elephantine limbs and probable ambling gait.[1]

[1] Mr. Charles Earle, after a careful revision of this family, has suggested that Cope's species should be greatly reduced in number, many of them in his opinion being only due to differences of age and sex. *Bathmodon*, *Metalophodon*, *Ectodon*, *Manteodon*, may all be included in the genus *Coryphodon*.

C. hamatus was about six feet long; while some species were no larger than a tapir, others were as big as an ox (see Plate XVIII.).

Professor Cope thinks that in general appearance the Coryphodons resembled the bear more than any other living animal, with the important exception that in their feet they were like elephants. The artist and author, however, in making the accompanying restoration, have been guided by the evident relationship between *Coryphodon* and the *Dinoceras* (described in our former work[1]). "The movements of the Coryphodons," says Professor Cope, "doubtless resembled those of the elephant in its shuffling and ambling gait, and may have been even more awkward from the inflexibility of the ankle. But in compensation for the probable lack of speed, these animals were most formidably armed with tusks. These weapons, particularly those of the upper jaw, were more robust than those of the carnivora, and generally more elongate." There is no evidence that they had a proboscis—in fact, it is practically impossible. We may suppose from the nature of the teeth, and from other evidence, that they were vegetable feeders, but not restricted to any particular class of food; to a large extent they were omnivorous, like the hogs of to-day.

It is a little difficult to follow the curious interpretation arrived at by Professor Osborne and Dr. Wortman that "the positions of the fore and hind feet were absolutely different," the former being like those of the elephant (where only the tips of the toes touch the ground); the latter, in their opinion, like those of a bear and spreading out to rest on the ground (plantigrade).[2] This is not borne out by Professor Marsh's recent figure. Much valuable material for the study of the anatomy of this primitive mammal was collected by these two gentlemen, in spite of the

[1] *Extinct Monsters*, p. 151 (new edit.).

[2] *Bulletin of the American Museum of Natural History*, vol. iv. (1872), p. 121.

many difficulties with which they had to contend, during their expedition, in the summer of 1891, into the Big Horn and Wind River regions, where the Wasatch strata are found.

The sketch-map shown in Fig. 62, will give the reader a general idea of the positions of the different geological "basins"[1] in Western North America, where their strata have been found to yield so many valuable relics of ancient Tertiary life.

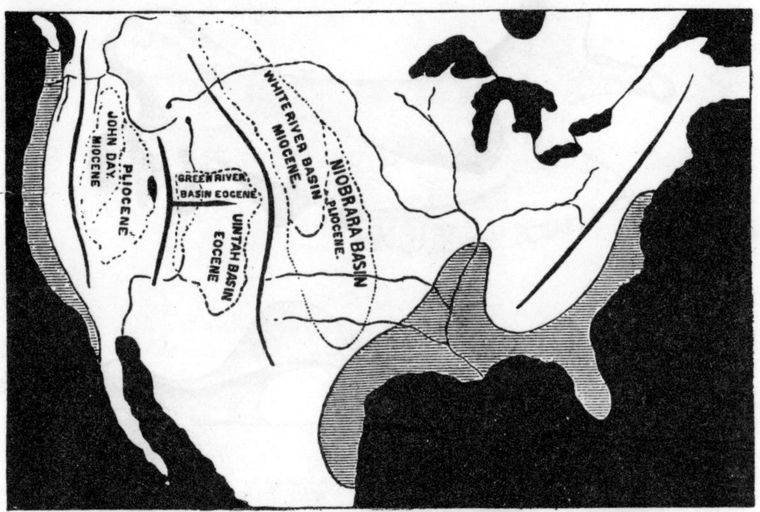

FIG. 62.—Map of North America in Tertiary times, showing roughly the outline of the coast, and sites of the principal fresh-water lakes.

The skull shown in Fig. 63, belongs to a huge American quadruped of Miocene age, the *Titanotherium*, which was one of the *Dinocerata* of Marsh. Probably *Brontops*, *Allops*, and *Menops* all really belong to this genus.

During the Miocene period there lived a very strange and primitive kind of elephant, known as the *Dinotherium*. Its head (four feet long, and three feet broad) was found at Eppelsheim, in

[1] When strata have been bent downwards into a kind of trough going far below the surface, they are said to lie in a "basin." Examples: the London basin, the Paris basin.

Hesse-Darmstadt, and is to be seen in the geological collection of the Natural History Museum, in a separate case (Glazed Case B, on plan). Some of the limb-bones may be seen not far off (Wall-case No. 39), together with a drawing intended to give an idea of the creature, as it looked when alive. The proportions of the head in this drawing do not agree with the actual specimen

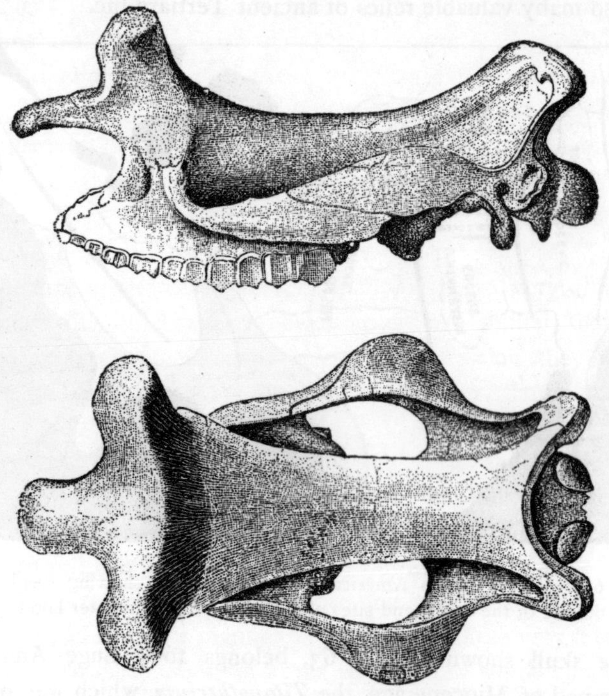

FIG. 63.—Skull of *Titanotherium*, from Eocene strata, North America. (After Marsh.)

close by: our artist's restoration in Plate XIX., which has been made after a careful examination of the head, will, we hope, give a more correct idea. Remains of this remarkable genus, which so puzzled palæontologists years ago, have been met with in various strata in South Germany, France, Greece (Pikermi), and

AN ANCIENT ELEPHANTINE MONSTER (DINOTHERIUM). MIOCENE PERIOD.

PLATE XIX.

Asia Minor. Closely allied forms occur in India, but none have yet been found in America.

The history of the discovery of this genus may be briefly given as follows. Isolated teeth were described in 1715 by Réaumer, in 1775 by Rozier, in 1785 by Kennedy. Cuvier ascribed a molar tooth to a gigantic Tapir (Tapir giantesque). Kaup, in 1829, established the genus from specimens found at Eppelsheim; at first, having only a lower jaw, he could not understand it, and not unnaturally put it upside down. But in 1835, Kaup and Klipstein found a complete skull at Eppelsheim (the one shown in Fig. 64). This head, when first exhibited in Paris, excited great interest among zoologists in that city. M. de Blainville read a paper on the subject before the French Academy of Sciences, pointing out what he considered to be its most remarkable points: he was wrong, however, in concluding that it was related (except, perhaps, in a very distant way) to the Manatees and Dugongs of the present day. Palæontologists can now form a very fair idea of the creature, because its bones have at different times been discovered. Kaup placed it between the tapir and rhinoceros. Dean Buckland compared the tusks to those of a walrus. In order to show how even the greatest anatomists may be deceived by mere fragmentary remains, like this skull and the few teeth at first known, it may be mentioned here that Owen at one time ascribed to this genus a jaw, with teeth, from Australia, which turned out afterwards to belong to the great extinct marsupial *Diprotodon!* (described in Chapter XI.). Several species of *Dinotherium* have been found, of various sizes. The skull shows that it possessed a fairly large proboscis. Unlike all other elephants, the two tusks are in the *lower* jaw, and curve downwards in the fashion of those of a walrus. The thigh-bone or femur (seen in Wall-case No. 39) is enough to show that the limbs were decidedly elephantine.

Dean Buckland, in his *Bridgwater Treatise*, states the conclusions he had arrived at with regard to the probable nature and habits of this remarkable animal; but we are not sure that he was right in supposing it to be of aquatic habits. He pointed out that it was mechanically impossible for a lower jaw nearly four feet long, and loaded with such heavy tusks, to have been otherwise than cumbrous and inconvenient to a quadruped living on dry land. For an animal destined to live in the water, such would, in his opinion, have been highly suitable and appropriate. The aquatic habits of the Tapirs, to which family the *Dinotherium* was more or less allied, seem to point to the conclusion that it

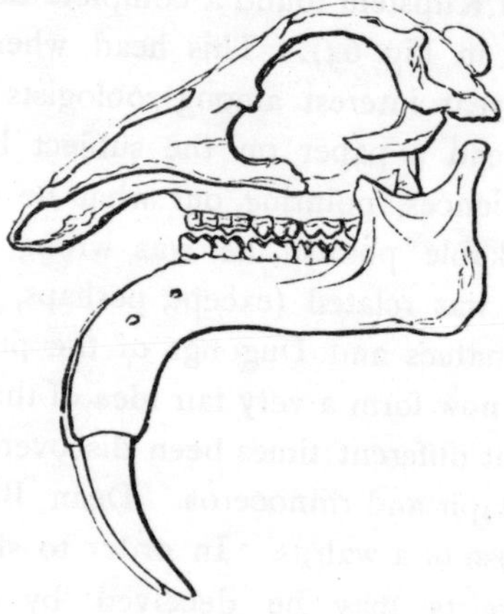

FIG. 64.—Skull of *Dinotherium giganteum*, from Eppelsheim. The lower jaw is a cast. An examination of the actual skull at the top will show it is considerably crushed in.

inhabited fresh-water lakes and rivers. To an animal of such habits the weight of the tusks, when sustained in the water, would not have been inconvenient. If we suppose them to have been employed as instruments for raking and grubbing up by the roots large aquatic plants from the bottom, they would, under such service, combine the mechanical powers of the pick-axe with those of the horse-harrow. The tusks may also have been applied with advantage to hook on the head of the animal to the bank, and perhaps, as in the case of the walrus, to assist in dragging the body out of the water. Some of these conclusions seem hardly justifiable, and we can see no difficulty in believing that the tusks were used to grub up the ground, or pull down branches of trees, as in the case of modern elephants.

A length of eighteen feet has been attributed to the creature; but this is only a probable estimate awaiting confirmation, or contradiction, as soon as a complete skeleton shall have been discovered. Its supposed aquatic habits seem to be in harmony with the geographical conditions of Europe during the Miocene period, for at that time there were many fresh-water lakes.

Other extinct elephantine monsters flourished in Europe in the same geological period with the *Dinotherium*, and continued to exist down to a later geological date; these were the Mastodons [1]—so named from the mammilated surfaces of their teeth.

Fig. 65 represents a restored skeleton of the *Mastodon angustidens*, from Miocene strata at Sansan, in France; and a restoration of the animal is shown in Plate XX. On comparing it with any modern elephant, one notices certain differences: to begin with, it possessed four tusks, two in the upper jaw and two in the lower; the legs are shorter, especially the hind pair. That it was provided with a proboscis admits of no doubt, and can be proved by several considerations. In the first place, the short neck shows that the mouth could not have reached the ground, and therefore such an organ was necessary in order to bring food to the mouth; secondly, the long tusks would equally make it impossible for the mouth to touch the ground; and thirdly, by the configuration of the skull, with its large surface for the attachment of the trunk. The teeth present certain differences when compared with those of an elephant. The office of the trunk, doubtless, was to seize and break off the boughs of trees for food, as in the modern elephant. It has been conjectured—though not with much reason—that the Mastodons

[1] Greek—*mastos*, teat; *odous*, *odontos*, tooth.

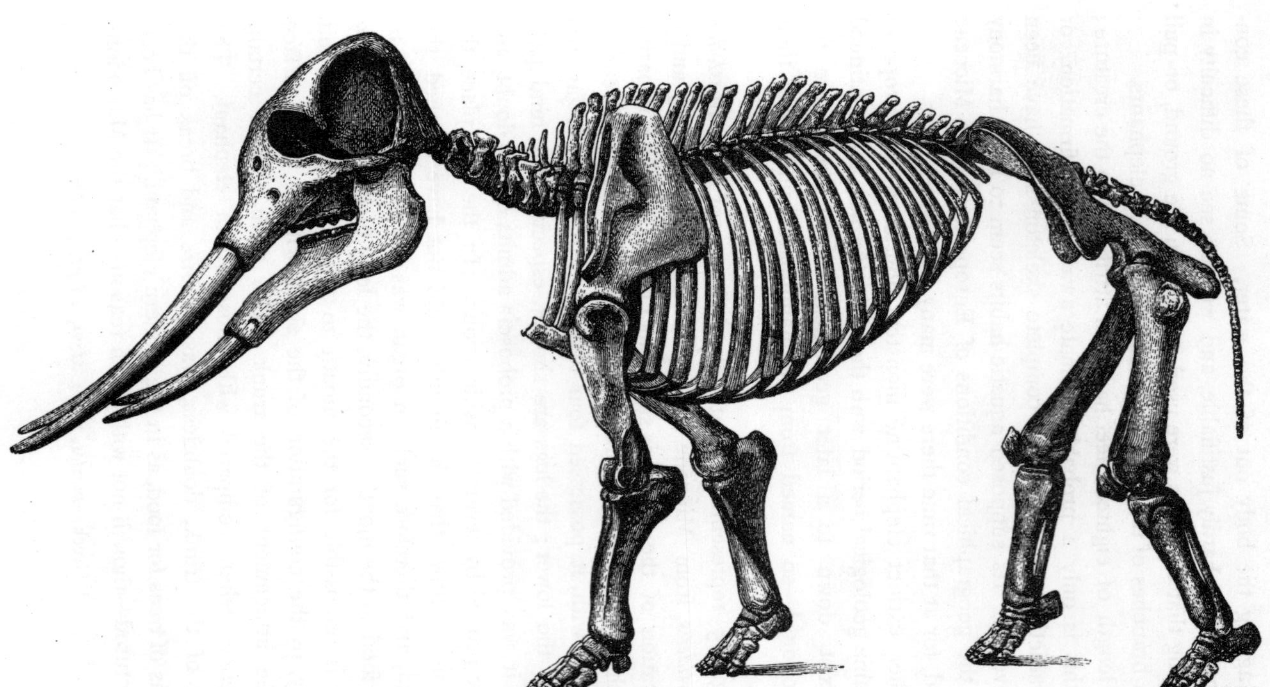

FIG. 65.—Skeleton of *Mastodon angustidens*, from Miocene strata. (After Gaudry.)

AN ELEPHANT WITH FOUR TUSKS (MASTODON ANGUSTIDENS). MIOCENE PERIOD.

PLATE XX.

were partly aquatic, and haunted the swamps of the Miocene period; but if so, it is not easy to understand the use of their tusks. A somewhat similar animal to the one above described, was the *Mastodon longirostris*, of which a skull was discovered at Eppelsheim in Hesse-Darmstadt. That the genus *Mastodon* lived on to a very recent period (possibly into the human period) is proved by the complete skeleton of *M. americanus* from Missouri, described in our former work.[1]

The Sivalik hills of Northern India have yielded a magnificent skull of one of the largest of the extinct elephants, the *E. ganesa*

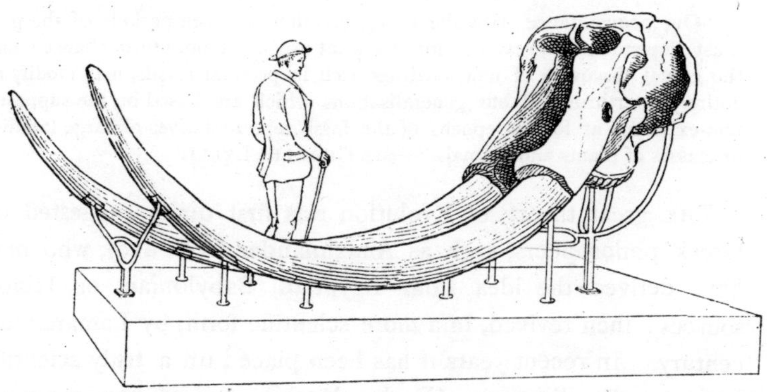

FIG. 66.—Skull and tusks of *Elephas ganesa*, from Older Pliocene strata, Sivalik Hills, North India. (From a drawing by J. R. Hutchinson.)

(see Fig. 66), the tusks of which measure ten feet six inches! This fine specimen was presented to the National Museum by General Sir William Erskine Baker. The power of these huge implements must have been truly enormous! The total length of the cranium and tusks is fourteen feet.

[1] *Extinct Monsters*, p. 220 (new edit.).

CHAPTER X.

HORSES AND THEIR ANCESTORS.

"Our acquaintance with the living creation of given periods of the past must depend in a great measure on what we commonly term chance; and the casual discovery of new localities, rich in peculiar fossils, may modify or entirely overthrow all our generalisations which are based on the supposed non-existence at former epochs of the fossil representatives of large families or classes of plants and animals."—Sir Charles Lyell.

The great theory of Evolution was first dimly suggested by Greek philosophers, such as Anaximander (B.C. 610), who may have derived the idea from Egyptian, Babylonian, or Hindu sources; then revived, in a more scientific form, by Lamarck last century. In recent years it has been placed on a truly scientific basis by the illustrious Charles Darwin, and is now generally accepted by naturalists and palæontologists. Indeed, it is hard to be a palæontologist in these days without being also an evolutionist—so abundant is the evidence derived from a study of extinct animals. Year by year the evidence is accumulating, and many workers in various parts of the world are discovering long-lost types which appear to link together some of the branches of the great Tree of Life. Marsh and Cope in America; Owen, Huxley, and others in England; have all been directing our ideas in the same course. At Pikermi in Attica; in the far Western States in America; at Sansans, Allier, Leberon and other localities in France, such great and important additions have been made of late to our knowledge, that it is now possible to make out certain lines of evolution in the Mammalia since their first important outburst at the beginning of the Tertiary period. We propose in the present chapter to trace, in the light of recent discoveries, the history of one important living group, as represented by the horse. The history of the horse and its various ancestors in Europe and America, as worked out by Huxley and Marsh, is undoubtedly one of the greatest triumphs of modern palæontology.

The series of fossil horses now known is so complete, that not a single important gap is left between the original five-toed ancestor, and the horse of to-day, with only one toe to each foot! Here, then, we have the most perfect evidence of the evolution of an animal from distant ages in the earth's history that has ever been presented to the world! Professor Cope's researches on the history of camels as illustrated from fossil remains have brought to light another line of evolution equally interesting if not quite so complete. According to the latter palæontologist, the very earliest ancestor of all the hoofed animals (and therefore of the horse) was the Eocene *Phenacodus.* We must therefore begin with this interesting and primitive little mammal, and some of its relatives, before we speak of the true horses. Only thus will the story be complete.

Professor Cope divides his suborder, the Condylarthra, into three groups or families, which take their names from three important genera, viz. *Periptychus*, *Phenacodus*, and *Meniscotherium.* Each of these strange types will now be briefly considered.

The first of them, *Periptychus*, was evidently very abundant during the earliest part of the Eocene period, when the famous Puerco beds were in the course of formation; for Professor Cope says that portions of fifty separate individuals (chiefly fragments of jaws) have come into his possession. It was apparently the largest and most specialised form of the family named after it. Three species were known when Professor

Cope wrote his papers on the above families (1884). He obtained a cast of a good part of the brain-case, and found that the part of its brain devoted to the sense of smell (olfactory lobes) was enormous, while the hemispheres of the higher brain (cerebrum) were small and flat. The cast disclosed quite the lowest type of a mammal's brain that has ever been brought to light. The discovery of this creature was an important event in the history of palæontology. The following description will perhaps serve to give the reader some slight idea of this remarkably primitive mammal, as pictured by Professor Cope. It had long legs and walked on the sole of its foot, as a bear does. Certain of the limb-bones suggest those of the elephant; but a fact like this does not necessarily imply any external resemblance to an elephant; elephants are a later branch from the same stem. The neck was remarkably short—another curious point of contact with the above order. *Periptychus* was a smaller animal than the *Phenacodus*, with a head of nearly the same size, and a long tail, stout at the base. There were no weapons of offence or defence, as far as one can see, and yet there is no saying what means it may have devised for those purposes. It is generally unsafe to hazard any speculations on that subject. Look, for instance, at the kangaroo: if it were an extinct animal, only known by its skeleton, the palæontologist might safely say it could excel in jumping and running; but he probably would never have guessed how effectively the creature can use its hind legs to kick or rip open a man. So with horses; they might seem at first rather helpless (except for their running powers), but see how a horse can kick when driven to bay! Still, the *Periptychus* had pretty large front teeth, and these, Professor Cope suggests, may have been capable of inflicting a severe bite. With regard to diet, we may suppose, from the nature of the teeth, that food was derived from both animal and vegetable sources.

All the known specimens of *Periptychus* were discovered by

Mr. David Baldwin, Professor Cope's assistant, in New Mexico; and not only these, but an immense number of fossil vertebrates of the Puerco age, are the results of his untiring and sometimes dangerous explorations. Few collectors can show such a record as his.

Leaving out certain other genera which are less well known, we pass on to the next family, of which *Phenacodus*[1] is the representative. The first specimen of this interesting animal—the most generalised mammal that has yet been discovered—

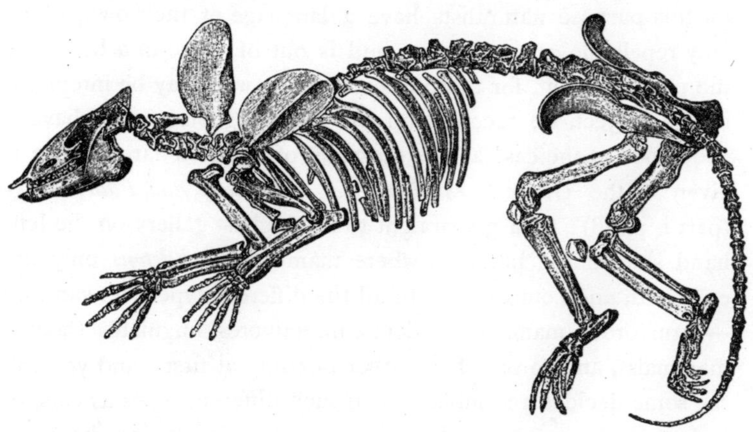

FIG. 67.—Skeleton of *Phenacodus primævus*, the oldest known ungulate. From Eocene strata, Wasatch, U.S. (After Cope.)

was dug out by Professor Cope himself from a bank of Eocene marl, on Bear River, Wyoming, and was found associated with *Coryphodon* and *Hyracotherium*. Other specimens were obtained, in less abundance, in strata of similar age in New Mexico. But the locality where the bones have chiefly been found is the Big Horn Basin of Northern Wyoming. Here Mr. Wortman has met with a great many specimens, and, amongst others, two almost entire skeletons of *P. primævus* (Fig. 67) and *P. vortmani*.

[1] Greek—*phenax*, a cheat; *odous*, tooth, because its teeth are deceptive.

These two species are somewhat different in size and build. Altogether nine species are more or less known.

The reader may obtain a very good idea of the skeleton of *P. primævus* from the beautiful cast exhibited in the Natural History Museum (Gallery of Fossil Mammals, Pier-case No. 9). Mr. Wortman brought two specimens from the Wind River Valley and two from the Big Horn Basin. It is, of course, quite impossible to describe the skeleton of any animal in popular language, much more that of an ancient creature belonging to an extinct order; for this purpose naturalists have a language of their own, which only repels the general reader, and is out of place in a book like the present. But, for those of our readers who may be interested in the subject, we recommend the following plan: first have a good look at the cast at the museum, of which a large figure is given in the *Guide to the Galleries of Geology and Palæontology* (part i. p. 28); then go straight to the highest gallery on the left-hand side of the building, where mammalian *skeletons* only are exhibited, and compare it with all the different types you can find —carnivorous mammals, rodents, insectivores, ungulates (hoofed mammals), and so on. It is rather puzzling at first; and you will see some decided resemblances to such different types as cats, or wolves, capybara among the rodents, and tapirs among the ungulates. This is a little bewildering; but when you have had a good look round you will almost certainly come to the conclusion that the nearest thing you can find is the skeleton of the American Tapir (*Tapirus terrestris*). Such was our own experience, and on reading Professor Cope's paper *afterwards*, we found that he points out several features in common between the skeletons of the little Eocene type and its modern relation, the common American tapir. Of course it was not so large or so heavily built, and the skull is different in several ways; but there is a striking resemblance. In size, our *Phenacodus primævus* was something between a sheep and a tapir.

It will readily be perceived that in the specimen above

ANCESTORS OF THE HORSE. EOCENE PERIOD.

Phenacodus. *Hyracotherium.*

PLATE XXI.

referred to the fore limbs have been somewhat displaced after the death of the animal, so that one shoulder blade appears much too high; but this can be allowed for. Beginning with the skull, we find that it is long, narrow, and rather pointed in front, where the nose was. The teeth are very peculiar, the molars or back teeth having several low rather conical protuberances, thus presenting a different appearance to those of a rhinoceros on the one hand, or of a deer on the other; but more like those of a pig, which comes of a very old race, and shows a good deal in common with the odd-toed ungulates—the rhinoceros, horse, and tapir. The feet are also of a primitive type in possessing five toes instead of one, as in the horse, or three, as in the tapir, or two, as in the deer. Evidently we have revealed here the primitive foot from which all the hoofed mammals are descended, as well as elephants, and probably even the carnivores and rodents!

A cast of the brain-cavity showed Professor Cope that the brain was decidedly primitive too, the hemispheres being very small, though they exhibited traces of convolutions. But *Coryphodon* had a brain still more primitive. One species, the *Phenacodus nuniensis*, was considerably larger than the others, being about the same size as a Malayan tapir. The others were small. *Phenacodus primævus*, Professor Cope thinks, had much the same proportions as the American tapir, and this has to some extent determined our restoration shown in Plate XXI.; but we must think of a *young* tapir, not a full-grown one. The middle three toes of both feet reached the ground, while the other two —one on each side—hung down without reaching the ground, like those of a pig. The tail was longer and heavier than that of any living hoofed animal, and reminds one more of the tail of a cat or a lion. The eyes were small, and the muzzle long, but very soft above. Whether there was a small proboscis it is hard to say, but there are some suggestions of such an appendage in the nasal bones. With regard to diet, it is pretty safe to conclude that the animal was not particular in its choice of food, and, like a pig, would eat almost anything that might come in its way.

We know that life is a struggle for all creatures—man not excepted; and so it must have been even in the dim and distant past of the Wasatch Eocene period. How, then, did *Phenacodus* maintain the struggle? It must have had enemies, though there were no lions or tigers then. It was not endowed with as much strength as a tapir; and, as far as the skeleton shows, it possessed no weapons, either of offence or defence. Professor Cope concludes that the creature, when danger was near, sought refuge in flight. It may have been capable, he thinks, of running at considerable speed. Such a conclusion is suggested by the limb-bones, which show rather well-marked places for the attachment of muscles.

Turning to the other species, *Phenacodus vortmani*, we find the limbs rather long and slender for an Eocene mammal. Professor Cope believes that we have in this genus the primitive ancestor of a great many mammals, and among others, of the horse. It may be, then, that we see here the first beginning of the development of that remarkable increase in running power exhibited in the ancestral horses (to be presently described), accompanied by a gradual decrease in the number of bones in the limbs, and other changes which made for speed. In size, our *Phenacodus vortmani* was about equal to a bull-dog; but the head is smaller and the neck rather shorter, and not nearly so robust. The limbs show about the same proportions to the body as those of a bull-dog, but the fore limbs are shorter. The feet are a little longer than in the last species; indeed, Professor Cope has some doubts as to whether this animal is even of the same genus as the last. Here is his description of it: "We can thus imagine the *Phenacodus vortmani* as an animal with the comparatively slender build of the bull-dog, with a neck and head proportioned more as in the racoon, and with the rump more elevated than the withers, as in the peccary. The feet resembled those of a tapir or rhinoceros, but had a pair of short toes on each side which did not reach the ground. To this add a tail much like a cat's in proportions, and the picture is complete. The diet of this animal was omnivorous, with a smaller proportion of animal food than the hogs, for instance, use. The food is more likely to have resembled that of the quadrumana (lemurs, monkeys, etc.). What means of defence this species had is not easily surmised, as the canine teeth and hoofs are not large."

The last family of the group now under consideration is represented by a fossil known as the *Meniscotherium*, of which only three species are known. It occurs abundantly in the same Wasatch Eocene strata. *M. chamense* was about as large as a fox, but with a very different physiognomy. The profile is curved, the muzzle short, and the eyes large. The body is not so slender as that of a fox, or of the *Phenacodus*, having the more robust proportions of a racoon. The fore and hind legs were rather short, and of equal length, so that the rump was flattened as in the dog. There was a large tail. In diet the creature was probably a vegetarian.

We pass on now to consider the true fossil horses, as worked out by Professors Huxley and Marsh. The modern horse, and that which was known to man in the days before history was written, we may regard as a product of the latest geological period—the Pleistocene. The development of the horse from a primitive five-toed ancestor seems to have taken place along two separate lines, one in Europe and one in America. The latter is the most complete; for it so happens in that country the physical conditions which prevailed throughout nearly the whole of the Tertiary Era were singularly favourable to the preservation of the skeletons of those creatures which lived on land. The various members of the horse tribe that roamed over North America all through these long ages of the past were specially numerous in what is now the Rocky Mountain region, and their remains are sealed up in the old lake-basins which then covered so much of the country.

The most ancient of these lakes—which extended over a considerable part of the present territories of Wyoming and Utah—continued to exist so long during the Eocene period that the sand and mud brought into it by the agency of rivers actually accumulated therein to a thickness of about two miles! Here is found one of the oldest direct ancestors of the horse, the *Orohippus* of Marsh. During the middle Tertiary, or Miocene period, two other lakes existed on either side of the great Eocene basin. The largest of these, to the east of the Rocky Mountains, extended over portions of what are now Dakota, Nebraska, and Colorado. The clays deposited in this lake form the "Bad Lands" of that region, so well-known for their fossil treasures. The other Miocene lake was west of the Blue Mountains, where eastern Oregon now is, but, having since been overflowed by a vast sheet of basalt, its thickness is unknown. In this basin the *Miohippus* of Marsh first makes its appearance.

During the later Tertiary, or Pliocene period, a vast development of the horse tribe took place, so that great numbers of these animals left their remains in the lake-deposits of that time. The largest of these lakes had the Rocky Mountains for its western border, and extended from Dakota to Texas, its northern part covering the bed of the older Miocene basin. Another Pliocene basin, of unknown limits, extended over the older Tertiary strata of Eastern Oregon, and evidence of yet others may be seen in Idaho, Nevada, and California. In all of these fossil horses have been found; but the most important localities are the region of the Niobrara River East of the mountains, and the valley of the John Day River in Oregon (see sketch-map, p. 191).

It will thus be seen how abundant is the material for tracing the evolution of the horse in America. In that country the *Equus Fraternus* of Leidy is believed to be almost, if not entirely identical with the *Equus caballus* (Linn.) of the old world. The earliest ancestor of the horse in the new world, at present known, is the *Eohippus*, of the Eocene period (*Phenacodus* and its allies may be direct ancestors; but, for our present purpose, we may set it aside as being too remote). Several species of *Eohippus*[1] have been found, all about the size of a fox. Like most of the early hoofed mammals, it possessed forty-four teeth, the molar teeth having short crowns. In the fore feet there were four well-developed toes and a trace or rudiment of another; the hind feet had three toes. Here we have clear evidence that the horse tribe had begun to separate themselves from the rest of the hoofed animals. The two bones of the fore-leg (radius and ulna) are quite distinct, as in *Phenacodus;* but we shall see presently how one of them gets less and less. The next stage in the progress of evolution, as represented by *Orohippus*,[2] marks only a very slight change, the rudiment of a fifth digit in the fore limb has gone, and we see only four. That rudiment represented a thumb; so now we have in the *Orohippus* only the second, third, fourth, and fifth digits. The reader will find these shown in Fig. 68, in the bottom row. Here the big digit represents the middle finger of our own hands, the one on the left of it being number two, and the two on the right numbers four and five, counting the thumb as number one. The next drawing in the bottom row shows the three digits of the hind limb, namely, a middle toe and one on each side of it, as in the living tapir. Then follow the radius and ulna, which are seen to be separate—the ulna being the long one. Then we see the tibia and fibula of the hind limb—the tibia being the thick one; and, lastly, we see an upper molar tooth (represented in two ways), and then a lower molar. These drawings, which were made under Professor Huxley's supervision, are quite sufficient to illustrate the changes that took place in the ancestors of the horse; but they do not show the increase in

[1] Greek—*eôs*, dawn; *hippos*, horse.

[2] Greek—*oros*, mountain; *hippos*, horse; because found in the Rocky Mountain region.

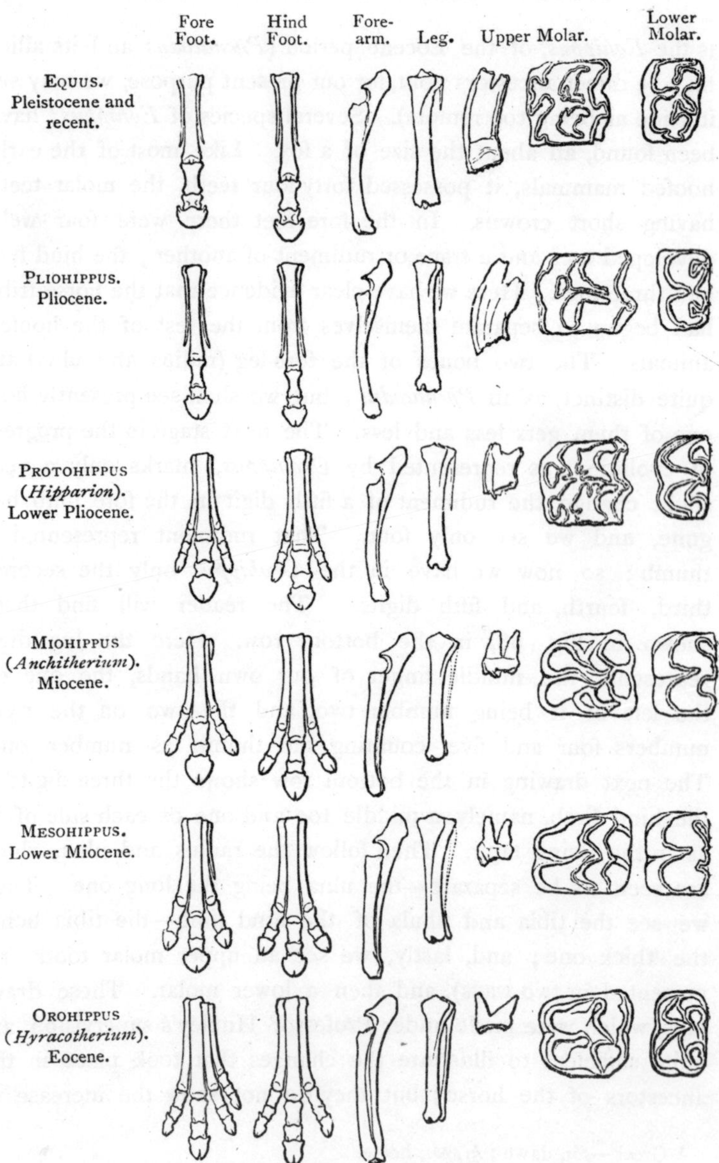

FIG. 68.—Limb-bones and teeth of American fossil horses (after Marsh), illustrating a gradual progression from Eocene times. The *Phenacodus* is omitted. The nearest European forms are given in *italics* underneath.

size that took place. *Orohippus* was found by Marsh in middle Eocene strata, in the same beds as the Dinoceras described in our former works; it was but little larger than *Eohippus*, and in most respects very similar. Now, the *Hyracotherium* of Europe and America came near to *Orohippus*, so that the delicate drawing on p. 210 (taken from Cope's great work on the Tertiary vertebrata) will give us a very good idea of this graceful little creature, which ought to be specially interesting as one of the earliest members of the horse tribe; and we trust that the restoration of it in Plate XXI. will help the reader to picture it to himself as it was when alive. It is somewhat too large in proportion to the *Phenacodus*, but if drawn much smaller perhaps some of its features might be lost. We have no guarantee for the stripes which the artist put on both the animals. The young American tapir has stripes like those represented on the body of our *Phenacodus*; but it is not likely that the tapir comes at all in the line of the fossil horses, nor is there any positive reason to suppose that *Hyracotherium* was striped: the artist thought the stripes would improve the picture—as they do,—but they were not due to any suggestion from the writer.

Mesohippus, from the base of the Miocene, is rather abundant in the beds in which the great *Brontotherium* was found. Here we see, from our diagram (Fig. 68), that the fifth digit has become decidedly smaller, and the fibula of the hind limb was probably incomplete. The next stage, represented by *Miohippus*, is an important one. This genus comes very near to the European form *Anchitherium*; and here we see that the small fifth digit of the *Mesohippus* has practically vanished, while the ulna has lost one end, and is beginning to dwindle down as shown in the later forms, such as *Hipparion*. There are several species of *Miohippus*, all larger than *Mesohippus*, and more specialised both in the skull and feet. They were about as large as a sheep. A still greater change manifests itself in the *Protohippus* of the lower Pliocene—which resembles most nearly the European form

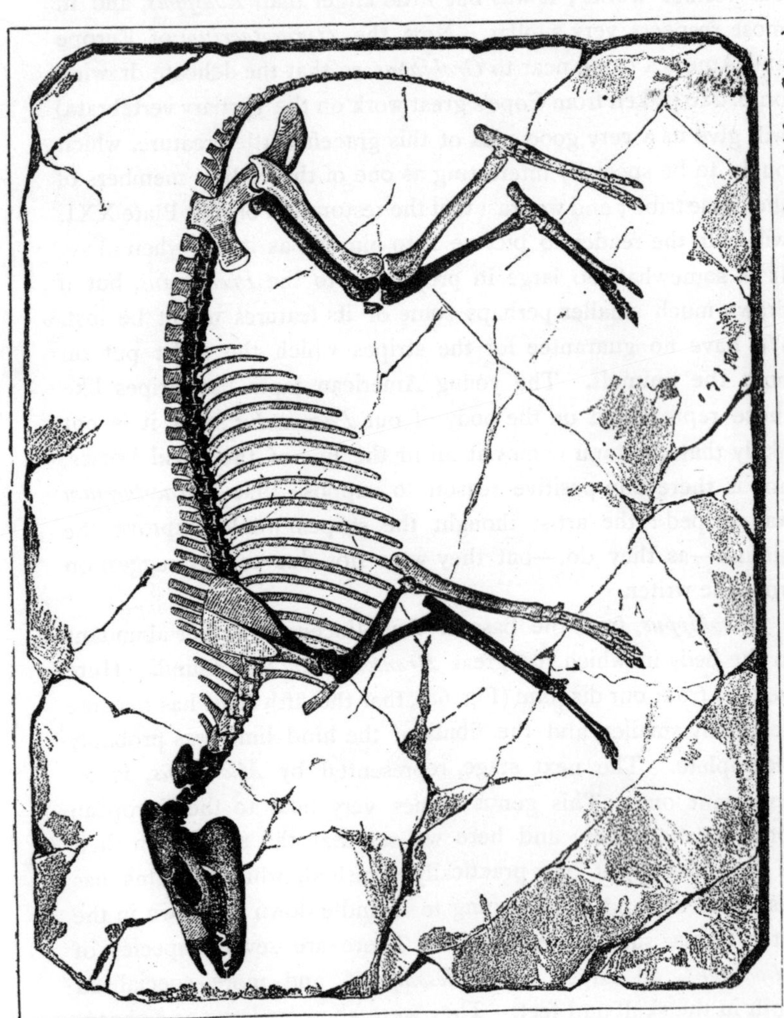

FIG. 69.—Skeleton of *Hyracotherium venticolum*, from Eocene strata, North America. (After Cope.)

Hipparion. Some of the species of *Protohippus* were as large as an ass, and the European *Hipparion* was quite as large; the latter we have ventured to restore as seen in Plate XXII., where it is somewhat too large in proportion to the wild horse drawn below—but this was done purposely, in order to show it better. Here (in *Hipparion*) we have another decided advance towards the modern horse; the two extra toes have become much smaller, so much so that we may suppose them to have been of no use at all, as suggested in our restoration. In other words, they have become "rudimentary."[1] In the fore limb of *Protohippus* the ulna has dwindled considerably, and the upper molar tooth is getting deeper, and its surface for grinding the food more complex; in fact, very like that of *Equus*, shown at the top of the diagram. In the Pliocene period we have the last stage of the series before reaching the horse itself, viz. the *Pliohippus*, which has lost the small hooflets of the previous American form, and the two side digits are reduced to mere "splints," as in a modern horse. This transition is a greater one than the last; but yet the gap between the two forms is not very great *Pliohippus* does not seem to have been any bigger than *Protohippus*; but we see at once how very horse-like it must have been.[2]

Only in the upper Pliocene deposits does the true horse appear, and then the genealogy is complete. It roamed over the whole of North and South America, and soon after *seems* to have become extinct. There is no doubt that man and the horse were contemporaneous in early days; but it can be proved beyond doubt that, at the time of the Spanish conquest, few if any horses were left. It has been thought, from certain

[1] It is important to note that in *Protohippus* the two side digits are larger than in *Hipparion*, and were probably of some use, whereas in the latter form they would seem to be too small to touch the ground.

[2] Professor Marsh has collected remains of some thirty or forty species of more or less horse-like animals; some of these, such as *Parahippus*, *Hyohippus*, and *Merychippus*, may not be in the line of descent.

references contained in old narratives, that at least in South America the animal may possibly have still lingered on after the coming of the Europeans. What cause can have led to its extermination, it is impossible to say. The present race of wild horses, which roam in such vast herds on the Pampas, are not the descendants of the fossil horse of South America, but have sprung from those introduced by the Spaniards more than three hundred years ago.

We pass on now to trace the line of descent in Europe. In the year 1839 Sir R. Owen described an imperfect skull of a small animal not larger than a fox, which was discovered in the London clay of Herne Bay, in Kent, under the name of *Hyracotherium*, because at that time it was (wrongly) supposed to be related to the living *Hyrax*—that peculiar little creature inhabiting Syria, which may be the "coney" of Scripture, and looks like a rabbit. Similar and more complete remains were afterwards found in various parts of Europe, and more abundantly in America (see Fig. 69). To a closely allied form the name of *Pachynolophus*[1] has been given, while *Pliolophus* and *Orohippus* are probably identical; and they are all nearly related to a previously known, but larger animal, called by Cuvier *Lophiodon*.

In Europe the horse is generally traced from the small and slender *Paloplotherium*, of which the whole skeleton is known, by way of *Pachynolophus* to the *Anchitherium*; thence to *Hipparion* and so to the *Equus caballus*.

Paloplotherium has been found in France, England, and elsewhere, and looks like *Hyracotherium*. *Hipparion* seems to have had a wide geographical range, for its remains have been discovered in the Sivalik hills of Northern India, in China, and at Maragha in Persia, in Samos, in Pikermi (Greece), in France, Germany, and England. The *Anchitherium* occurs in Miocene

[1] Under the name *Hyracotherium* some very different forms have been confounded; and Owen's species may possibly be identical with Cope's *Phenacodus*.

THE HIPPARION (UPPER PICTURE), AND A WILD HORSE (LOWER PICTURE).
PLATE XXII.

strata at Sanson in France, and another small species comes from the Miocene White River beds of Nebraska in North America.

Equus stenonis of the Pliocene period is regarded as the immediate ancestor of the horse in the Old World.

And here a word of caution seems to be necessary. When we find fossil types which appear to be intermediate between some that are living at the present day, it does not follow that they are *really* intermediate. Thus, as already pointed out, the ancient *Palæotherium* has much in common with the horse on the one hand, and with the rhinoceros on the other; but it does not therefore represent the form through which rhinoceroses have passed to become horses, or *vice versâ*. It probably means that the horse, the rhinoceros, and the *Palæotherium* are all the descendants of a common ancestor. We must, therefore, distinguish clearly between these "intercalary" types, as they are now called, and those "linear" types which more or less truly indicate the steps by which the transition from one group to another was effected. The changes and modifications of the limbs of the horse's ancestors were, of course, only acquired very gradually—as, indeed, is indicated by the different periods of the Tertiary era during which they lived; and it is obvious that they all brought about increased speed in running, especially over firm and unyielding ground.

Speaking of these changes, Sir William Flower says,[1] "Short, stout legs and broad feet, with numerous toes, spreading apart from each other when the weight of the creature is borne on them, are sufficiently well adapted for plodding deliberately over marshy and yielding surfaces, and the tapir and the rhinoceros, which in the structure of their limbs have altered but little from the primitive Eocene forms, still haunt the borders of streams and lakes and the shady depths of forests, as was probably the habit of their ancient representatives; while the horses are all inhabitants of the open plains, for life upon which their whole organization

[1] *The Horse: A Study in Natural History.* (London, 1891.)

is in the most eminent degree adapted. The length and mobility of the neck, position of the eye and ear, and great development of the organ of smell, give them ample means of becoming aware of the approach of enemies; while the length of their limbs, the angles the different segments form with each other, and especially the combination of firmness, stability, and lightness in the reduction of all the toes to a single one, upon which the whole weight of the body and all the muscular power are concentrated, give them speed and endurance surpassing that of almost any other animal."[1] At the same time that these changes were taking place (and in accordance with the "Law of Correlation") certain important changes in the structure and mode of growth of the teeth were being developed, the consequence of which was that they were able to masticate and grind up their food much more perfectly, and thus, no doubt, to obtain therefrom more vital energy with which to supply their muscles and nerves for working their limbs more rapidly as they became fitted to attain higher speed. They gained a grinding tooth for each toe that they lost. Of these ancient horses it may well be said they "ran the race that was set before them" or, perhaps we might say, which they set before themselves; and the victory was to the quickest runners—those who could "beat the record" made by their predecessors! It is a wonderful story, this, of the evolution of that noble and useful animal to which man is so much attached; and one cannot help sometimes wondering whether to some extent the *will* of an animal may not be an important factor in Evolution, although it is the fashion to ignore it, and to attribute organic changes to Natural Selection, or the "survival of the fittest." Mind has a powerful influence over matter, and can we not conceive that the earliest ancestors of the

[1] Visitors to the Natural History Museum will do well to examine the admirable set of casts and actual bones of fossil horses in Gallery No. 1. Complete casts of *Phenacodus* and *Hyracotherium* are there exhibited. In the central hall may be seen the skeleton of a man and of a horse side by side, in the same relative position.

horse, such as *Phenacodus*, finding (perhaps for the first time in the world's history) boundless grassy plains before them, were impelled by a strong desire to run? The exercise itself must have been enjoyable and invigorating; besides, the more ground they could cover in a day the greater choice of pasture they would find, while at the same time increased swiftness of foot meant greater safety from flesh-eating enemies. Nature would therefore encourage such a wish, and give success to the swiftest. Not that we have any doubts as to the wonderful results which Natural Selection can produce; but we cannot help thinking that the animals themselves may have made a *determined effort in the direction of attaining speed*—instead of allowing everything to be done for them. Why should they not—of course with the co-operation of Nature—have "worked out their own salvation"?

It is a well-known fact that young horses are occasionally born with small extra toes on their feet. Suetonius[1] has given an account of the famous steed of Julius Cæsar. According to the historian, Cæsar "used to ride a remarkable horse, which had feet that were almost human, the hoofs being cleft like toes. It was born in his own stables, and, as the soothsayers declared that it showed its owner would be lord of the world, he reared it with great care, and was the first to mount it; it would allow no other rider." This account clearly indicates the presence of extra digits, and the occurrence is more frequent than is generally supposed. Professor Marsh has made a study of this subject, since it has an important bearing upon the history of the evolution of the horse as given above; for in the light of modern teaching such occurrences are but "reversions" to older types that lived ages ago. He has therefore examined a large number of living animals with this peculiarity, and several interesting specimens of the same character have been sent to him: he has likewise received photographs, drawings, and detailed descriptions of

[1] *De Vitâ Cæsaris*, lxvi.

various other examples, the authenticity of which cannot be questioned. The various cases he has come across may be summed up as follows :—

Sometimes there is only one extra digit on one foot. This is always much smaller than the main or third digit, the largest he has seen being about one-half its size, and the smallest very diminutive. This extra toe is almost invariably on the inner side of the main digit, and usually on the fore foot. Sometimes it is entirely hidden beneath the skin, the only external evidence of its presence being a prominence, which, on close examination, may be found to contain, below the "splint-bone," two or more bones corresponding to the finger-joints of our own hands. Sometimes a corresponding extra toe is developed on the other fore foot. At other times a second extra digit has been found, as well as those above mentioned, which may be either smaller or larger.

Coming now to the hind feet, we find that an extra toe may sometimes be present on one or both of them, and always on the inside. As a rule, however, these posterior toes are much smaller, and often concealed beneath the skin. Fig. 70 is a drawing of a horse from Cuba, in which the extra toe is well developed on all four limbs. This animal was examined by Professor Marsh during its lifetime.

Occasionally the hind feet have two extra toes, while the fore feet have only one, as in the horned horse from Texas, figured by Marsh;[1] but in rare cases both the fore and hind feet may each have two extra toes fairly well developed, and all nearly of the same size. In such cases the result is something very like the feet of the ancient American type *Protohippus*, and the European *Hipparion*. These extra digits are, of course, the second and fourth represented in our diagram on p. 208.

In the museum at Yale College there is preserved a specimen of a horse which, besides having two extra toes on each foot, had

[1] *American Journal of Science*, vol. xliii. p. 344.

concealed beneath the skin the remains of another one corresponding to the human thumb. In such a rare case only the little finger is wanting to bring us back to something like the hand of the most primitive Eocene types, such as the *Phenacodus!* Professor Marsh examined a horse of this kind when it was alive, and, at its death, the owner, Mr. Theodore F. Wood, of New Jersey,

FIG. 70.—Eight-toed Cuban horse. (After Marsh.)

presented it to the museum of Yale College. The animal was widely known to the public as "Clique, the horse with six feet," having been exhibited for many years both in America and in Europe. It was said to be from Texas. The extra digit representing the thumb in the human hand was not visible externally, but could be detected beneath the skin.

Most of the horses with extra toes which have been noticed in America have been raised in the South-west, or from ancestors bred there, so that their connection with the mustangs, or half-wild stock of that region, becomes more probable. It is well known that when animals run wild they are more likely to "revert" to ancestral types, and the late ancestors of the mustangs were in a wild state for several hundred years.

We have pointed out in this and the previous chapter that the mammals of early Tertiary days had small brains, and the oldest ancestors of the horse are no exception to the rule. Professor Marsh has clearly proved that during this era a gradual increase in the size of the brain took place. It is interesting to find that the growth was mainly confined to the cerebral hemispheres, or higher portion of the brain. In most groups of mammals the brain has gradually become more convoluted, and thus increased in quality, as well as in quantity. In the long struggle for existence during the whole of Tertiary times the big brains won, as they do now. Applying this to our ancestral horses, it is easy to see that, as they acquired greater speed, and so roamed over larger tracts of country, they found a greater variety of objects on which to exercise their wits, besides the necessity for much observation and reflection in finding their way across country.

CHAPTER XI.

EXTINCT WHALES AND WOMBATS.

"The earth from her deep foundations unites with the celestial orbs that roll through boundless space, to declare the glory and show forth the praise of their common Author and Preserver; and the voice of Natural Religion accords harmoniously with the testimonies of Revelation, in ascribing the origin of the Universe to the will of One eternal, and dominant Intelligence, the Almighty Lord and supreme First Cause of all things that subsist, 'the same yesterday, to-day, and for ever.'"—DEAN BUCKLAND.

It would not be safe to venture any positive statement or opinion as to when the great cetacean order of mammals, which includes such creatures as whales, dolphins, and porpoises, first appeared on the earth. But as far as the geological record is known at present, it would seem as if they were ushered in at the commencement of the Tertiary era, at the same time with many other new forms of life. The original ancestor of all these marine mammals may, for all we know, have been evolved some time before then; say in the Cretaceous period, or perhaps during that great and partly unknown interval between the Secondary and Tertiary eras. What we do know for certain is, that fossil evidence of their existence first comes to light in the Eocene strata of North America.

The oldest known whale is the curious *Zeuglodon*, from the marine Eocene strata of Alabama. At Claiborne, according to Sir Charles Lyell, there occur numerous species of shells, besides many sea-urchins (echinoderms) and abundance of sharks' teeth.

But the most remarkable remains found there are those of the extinct whale, known as *Zeuglodon cetoides*,[1] and so called from the yoke-like form of its double-fanged molar teeth, which are six inches in length. It has also been found in Mississippi, Georgia, and South Carolina. No strata of later date than the Eocene period contain its remains, as far as we know; and it has never been found out of the Northern Hemisphere. *Zeuglodon* must certainly have been a very remarkable creature, and it is to be hoped that palæontologists will some day meet with a complete skeleton, such as may afford the material for drawing its entire bony structure. Unfortunately, although a great many finds have been made in Alabama, no complete skeleton has yet been got together. The remains of at least forty individuals have been met with, so that we may venture to say that this ancient cetacean once flourished vigorously. So numerous are the vertebræ in the above-mentioned locality that it is said they are used for making fences, and that farmers even burn the bones in order to get rid of them! How little they know that hundreds of museum curators all over the world would be only too glad to procure some of this "rubbish"! With the spread of education, and an increasing desire to learn a little about the former history of our world, let us hope that in time agriculturists will endeavour to co-operate with men of science in rescuing from destruction the fossilised remains of the earth's former inhabitants.

Sir R. Owen declares the *Zeuglodon* to have been "one of the most extraordinary of the Mammalia which the revolutions of the globe have blotted out of the number of existing beings." Some of the vertebræ found in Alabama have a length of eighteen inches, and a diameter of twelve inches, while the whole of the backbone, or vertebral column, has been found undisturbed.

Sir Charles Lyell says: "The vertebral column of one skeleton found by Dr. Buckley at a spot visited by me, extended to a

[1] Greek—*zeuglê*, yoke; *odous*, *odontos*, tooth.

length of nearly seventy feet, and, not far off, part of another backbone, nearly fifty feet long, was dug up. I obtained evidence, during a short excursion, of so many localities of this fossil animal within a distance of ten miles, as to lead me to conclude that they must have belonged to at least forty individuals."

This animal is particularly interesting as marking the first appearance of the very distinct order of cetaceans, so unlike any other order, except the Sirenia (Manatee and Dugong). It may interest the reader to know that the teeth of this carnivorous whale were first described and figured by the mediæval writer Scilla, in his treatise entitled *De Corporibus Marinis*. The original specimens were obtained in Miocene strata at Malta, and are now preserved in the Woodwardian Museum at Cambridge. (Excellent casts are exhibited in the Natural History Museum. Wall-case 22, Gallery No. 1.)

Professor D'Arcy Thompson thinks that *Zeuglodon* and other members of its family are related to the seals, and have no direct affinities with the cetacea; their teeth certainly are very similar, but unfortunately we do not at present know what their limbs were like—though one would expect, from the long pointed head of *Zeuglodon*, that the body also was long like a whale's, and adapted for rapid progress through the water, rather than for progression on land. One cannot easily picture a creature with such a skull walking on the land, even in the awkward manner of a seal; but we must wait for further evidence on this point. A portion of a skull of *Zeuglodon* was found in the Barton Clay, an Eocene deposit in Hampshire.[1]

We have already recorded one or two cases in which curious mistakes have been made by palæontologists; but the student soon finds how many and great are the difficulties that have to be overcome by those who study lost forms of life, difficulties which

[1] Professor Dames, of Berlin, has recently described certain bony plates found with Zeuglodont remains, which probably indicate that its back was protected by an armour of scutes (see *Natural Science*, vol. iv. p. 175).

in most cases are due to the imperfection of the material at their command, or the bad state of preservation in which the specimens are often found. The old saying, "Humanum est errare," is particularly true of palæontological workers. But, at least, it can be said of them that they are honest, and never attempt to delude the public. There have, however, occasionally arisen enterprising persons, more desirous of making money than of furthering scientific truth, who have, for a time, succeeded in the grossest deceptions. Matzuyer, about 1613, exhibited some Mastodon bones as those of a human giant (*Teutoboccus rex*) and made a great stir in France; and now we have to record another case in connection with *Zeuglodon.*

In the year 1845, Dr. Albert Koch exhibited a large skeleton of a fossil animal, under the name *Hydrarchos Sillimani*, the former of which would mean "King of the Sea," while the latter seemed to indicate the approval of Professor Silliman. These remains consisted of a head and vertebral column, measuring in all one hundred and fourteen feet, a few ribs and parts of some supposed paddles. Dr. Koch made the public believe that all the bones had been found together in such a way as to prove that they belonged to a single individual, and that the numerous vertebræ formed a connected series. This, however, was a mere fabrication, as was shown by Professor Wyman, who says "these remains never belonged to one and the same individual," and that the anatomical characters of the teeth prove them to be, not those of a reptile, but of a warm-blooded mammal. He came to the conclusion that the greater part of the bones belonged to the genus *Basilosaurus* (King-saurian) of Harlan, a name which the American geologists still retain for *Zeuglodon.* This skeleton represented *at least* two separate individuals.

The following letter on the subject, from the well-known geologist, Dr. G. Mantell, appeared in *The Illustrated London News* of November 4, 1848 :—

"SIR,

"Will you allow me to correct a statement that appeared in the last number of your interesting publication? The fossil mentioned at the conclusion of the admirable notice of the so-called sea-serpent, as having been exhibited in America under the name *Hydrarchos Sillimani*, was constructed by the exhibitor, Koch, from bones collected in various parts of Alabama, and which belonged to several individual skeletons of an extinct marine cetacean, termed *Basilosaurus* by the American naturalists, and better known in this country by that of *Zeuglodon*, a term signifying yoked teeth. Mr. Koch is the person who, a few years ago, had a fine collection of fossil bones of elephants and mastodons, out of which he made up an enormous skeleton and exhibited it in the Egyptian Hall, Piccadilly, under the name *Missourium*. This collection was purchased by the trustees of the British Museum, and from it were selected the bones which now constitute the matchless skeleton of a mastodon in our National Gallery of organic remains [then in the British Museum]. Not content with the interest which the fossils he collected in various parts of the United States possess, Mr. Koch, with the view of exciting the curiosity of the ignorant multitude, strung together all the vertebræ he could obtain of *Basilosaurus*, and arranged them in a serpentine form, manufactured a skull and claws, and exhibited the monster as a fossil sea-serpent, under the above-mentioned name, *Hydrarchos*. But the trick was immediately exposed by the American naturalists, and the true nature of the fossil bones pointed out.[1]

"Bones of the *Basilosaurus* have been found in many parts of Alabama and South Carolina, in greensand belonging to a very ancient Tertiary formation; hundreds of vertebræ, bones of the extremities, portions of the cranium and of the jaws with teeth have from time to time been collected. Remains of

[1] After this, Koch's sea-serpent was carried to Dresden.

species of the same genus have also been found near Bordeaux and in Malta.

"Professor Owen has shown that the original animal was a marine cetacean, holding an intermediate position between the Cachelots and the herbivorous species. It must have attained a length equal to that of the largest living whales, for a series of vertebræ was observed *in situ* that extended in a line sixty-five feet. . . .

"GIDEON ALGERNON MANTELL.

"19, Chester Square, Pimlico, Oct. 31, 1848."

Squalodon (see Fig. 71) represents another extinct genus of cetaceans, of which very little is known beyond the teeth and

FIG. 71.—Head of *Squalodon*, from Eocene strata, Alabama.

skull. It was formerly classified with *Zeuglodon*, but its skull is more like that of a dolphin, although the teeth are somewhat like those of the former. Teeth and fragments of skulls of *Squalodon* have been frequently found in the marine European deposits of Miocene age—especially in the Vienna basin, many parts of France, and the Crag formations of Antwerp and of Suffolk; also in strata of corresponding age in North America and South Australia. A few isolated teeth occur in the cave-deposits of Italy. The Natural History Museum contains an excellent coloured reproduction of the skull of *Squalodon* (Pier-case 21, Gallery No. 1).

Remains of the gigantic Sperm Whale or Cachalot (*Physeter macrocephalus*) are found in the Forest-bed of Cromer, and also in a later deposit in South America. A number of Miocene and Pliocene forms allied to the Cachalot have been described under various names. Remains of the curious Narwhal, with its long spear-like tusk, have been found in the Norfolk forest-bed. Professor Owen has described the skull of a thick-toothed Grampus that was found beneath the turf in the great fen of Lincolnshire, in the year 1843 (see Fig. 72). It belongs to the existing *Pseudorca crassidens*, but Owen thought it differed from the present dolphins of our own coasts. Extinct genera of dolphins are known to occur in some of the later Tertiary deposits of Europe and America.

That whales were contemporary with the mammoth in this

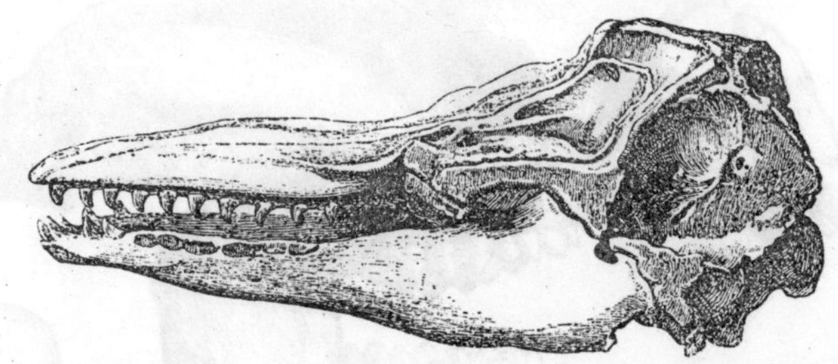

FIG. 72.—Skull of thick-toothed Grampus, *Pseudorca crassidens* (Owen). Fen, Lincolnshire.

country is proved by an interesting discovery made in the year 1828, in a cliff near Brighton, under the following circumstances. On the face of the cliff, in the ancient shingle which lies immediately upon the chalk, some fisherman had observed a huge bone that had been laid bare by an unusually high tide, and now projected two or three feet beyond the face of the cliff. Unable to remove it, they broke off the extremity, and sent a fragment of it to Dr. Mantell, who went to the spot a few days afterwards, and found that they had demolished a considerable portion of the bone in their attempts to remove it from its bed, and had only desisted from fear of being buried beneath the overhanging cliff, which is composed of loose materials. At last, and not without some danger, the entire specimen was left exposed. It proved to be the front nine feet of the left branch of the lower jaw of a whale-bone whale. On attempting to remove the specimen it broke into a thousand pieces. So, as the tide was approaching, this interesting find had to be abandoned. It must have belonged to a whale from sixty to seventy feet long.

Some of the ancient kangaroos of Australia, that lived during the Pleistocene period, were very much larger than those we

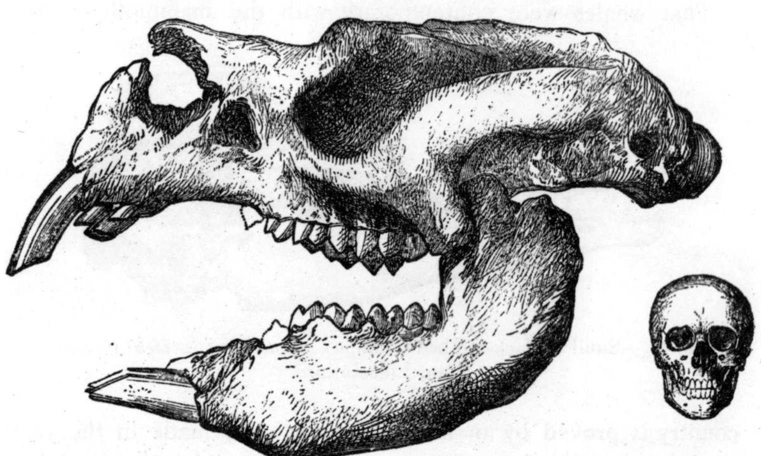

FIG. 73.—Skull of *Diprotodon*, a kind of Wombat, from Pleistocene strata, Australia. A man's skull placed for comparison.

see now in that country. One of the most remarkable of the extinct marsupials of Australia was the *Diprotodon*, of which a skull, measuring nearly three feet in length, may be seen in the Natural History Museum (Wall-case 27, Gallery No. 1), and it was so named on account of its two front teeth (see Fig. 73). The creature itself was probably even larger than a rhinoceros of the present day (standing six feet high), and is believed to have resembled a modern Wombat. The history of the gradual discovery of this strange marsupial may now be briefly given, since it serves to show how little ground there is for the popular belief that a Cuvier or an Owen could "restore" a whole animal from a jaw, or even a single tooth! Sir R. Owen himself was obliged to admit that his first conclusion, drawn from part of a jaw of *Diprotodon*, was quite wrong, as we shall see from the following account.

The name [1] was given by Owen to part of a lower jaw of the creature, obtained by Sir Thomas Mitchell in the Wellington Caves, New South Wales. Five years later, a drawing of part of a jaw with teeth reached England from the same source, and this Sir R. Owen believed to represent a kind of *Dinotherium*, indicating, for the first time, the occurrence of primitive elephants in Australia! In the same year, a portion of a molar tooth, associated with part of a thigh-bone and other fragments, was also received, and Owen wrote: "The fossils which my friend has now transmitted incontestibly establish the former existence of a huge proboscidian pachyderm in the Australian continent, referable either to the genus *Mastodon* or *Dinotherium* (!)."

Only a year later, however (1844), this prophecy proved to be incorrect; and, within a short time, Sir R. Owen was able, not merely to describe correctly most of the *Diprotodon's* skeleton, but also to distinguish another allied genius, the *Nototherium*. The feet alone remained unknown, and part of these were described as the toes of that fabulous monster, the "Great horned Lizard of Australia"!

In 1863 our late distinguished friend, Dr. George Bennet, wrote from Sydney to Owen, saying that his son, then in Queensland, had written home to say that he had found a place where a whole skeleton lay: the bones were immense; the head, he said, had been sent to Sydney some years previously (this may be the one now in the Natural History Museum).

[1] Greek—*dis*, twice; *protos*, first; *odous*, tooth.

Owen wrote to Sir Henry Parkes, in 1867, suggesting that his administration should send an exploring party to the limestone caves in Wellington Valley, discovered in 1832, and from which fossil remains were brought sufficient to show that the marsupial type of mammal formerly prevailed in that country as it does now. Consequently, a vote of two hundred pounds was passed, and a most valuable collection of fossils obtained.

The following passage, from a now historic paper by Sir R.

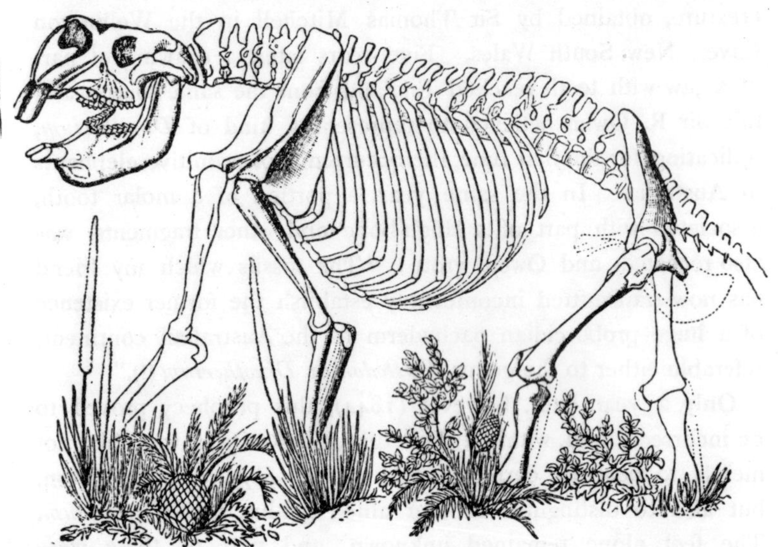

FIG. 74.—Skeleton of a huge extinct marsupial, *Diprotodon australis* (restored after Owen).

Owen, will serve to show with how great enthusiasm he pursued his most fruitful researches in palæontology. "Of no existing animal," he says, "of which a passing glimpse, as it were, had thus been caught, did I ever feel more eager to acquire fuller knowledge than of this huge Marsupial. No chase can equal the excitement of that in which, bit by bit, and year after year, one captures the elements for reconstructing the entire creature of which a single tooth or fragmentary bone may have initiated the quest; in the course of which one fully realises, with more or less exactitude, the picture which the laws of correlation had led one to frame of an animal which may have passed out of existence long ages ago." [1] Owen's restoration, published in the above paper, is shown in Fig. 74. But, only this last year (1893), a wonderful find has been reported from Australia, which will, doubtless, soon lead palæontologists to a complete and thorough knowledge of this great and mysterious marsupial. The first announcement was a brief telegram in *The Times;* but, later on, a letter appeared in *The Scotsman*, of September 14th, from Mr. William Kinmont. In the same enterprising newspaper we found, a few days afterwards (September 25th), extracts from an Australian account of the matter, the *South Australian Register* of August 12th. It seems that only a small portion of the area known to contain fossil remains has as yet been explored, and there is little doubt that a complete examination will yield results even surpassing those already made.

Mr. Kinmont writes as follows :—

Birksgate, Glenosmond, Adelaide, S.A., August 9, 1893.

"SIR,

"Some two months ago a discovery of mammoth fossil bones was made at Lake Mulligan in South Australia, which promises to be of incalculable value to science, and the 'record' discovery of the kind for the century. As the subject is one of world-wide interest, I propose to give you in brief, from my own knowledge and observation, the facts and particulars of the discovery.

"Lake Mulligan is situated in the central district of South Australia, and a more apparently uninteresting and forsaken spot than this locality could not be found in the colony. When the famous Australian explorers, Captain Sturt and Mr. John

[1] *Philosophical Transactions of the Royal Society*, vol. 160 (1870), p. 519.

M'Douall Stuart, went north-west from Menindie on the river Darling, in 1844, they came into a great stony desert lying to the eastward of Lake Frome, and the party endured the most terrible hardships from the frightfully barren nature of the country, and the extraordinary heat of the weather. . . .

"What a different picture of the past history of this country is brought to light by the recent discoveries! On the sides of these mountains lying between Lake Frome and Lake Torrens must have been grown huge trees, and all around there must have been a dense tropical growth, exceeding in luxuriance the forests of the eastern slopes of the Andes in South America. Every form of life was at that period huge and uncouth. On the grassy plains at the foot of the mountains roamed the *Diprotodon*, a gigantic animal allied to the kangaroo in its structure and habits, but of the size of an elephant.

"But to return to the recent discovery of the actual bones. The South Australian Museum, represented by the honorary director, Dr. E. C. Stirling, C.M.G., F.R.S., took the matter in hand, and despatched a party of competent men to excavate. These men have now been at work for some four months removing the gravel that buries the bones, and the discovery has been proved to be even greater than was anticipated, and already many new discoveries of the first importance have been made. For the first time in the history of palæontological science the complete skeleton of *Diprotodon australis* has been made known—this being a huge marsupial considerably exceeding the rhinoceros in size. The bones of this magnificent treasure, in the interests of science, are now in the Adelaide Museum. Remains also of a giant wombat as large as a half-grown bullock, of two or three kinds of gigantic birds equalling the *Moa* of New Zealand in size, several species of colossal kangaroos, as well as the head of what is undoubtedly a new species of *Diprotodon*, have been discovered, excavated, and brought to Adelaide. Over seventy different extinct animals and birds, hitherto unknown, are represented in the two thousand bones which have been dug out of the valley at Lake Mulligan.

"There can, in fact, be no doubt but that this discovery is the most remarkable one of its kind that has ever been made in Australia, and perhaps in the world. The South Australian Museum, with its limited funds, has done what it could, but with the work only fairly commenced the directors found themselves in the position of having to abandon the search for lack of means to go on. Sir Thomas Elder, however, the distinguished initiator and promoter of Australian exploration expeditions, with his

FIG. 75.—Restoration of *Diprotodon australis*.

usual liberality to the cause of science, and with a desire to advance the scientific reputation of South Australian work, has come to the assistance of the museum in this matter, and has enabled the directors to continue this most important and extensive work. Dr. Stirling proposes to start for Lake Mulligan himself this week, with a view of personally superintending operations; and there can be little doubt but that work in the future will yield results even surpassing those already made.

"I am, etc.,

"WILLIAM KINMONT."

The following is part of the article above alluded to from the *South Australian Register* :—

"Sir Thomas Elder, G.C.M.G., has once more placed South Australia under deep obligations to him by coming forward with the monetary help necessary to enable further explorations to be made of the area at Lake Mulligan, in which immense quantities of fossil remains of extinct animals and birds are known to exist. The generous assistance thus rendered has come at the nick of time. The authorities of the South Australian Museum, recognising the importance of the discovery in the interests of science, have already done much to secure specimens of the bones which have been preserved from destruction; but the funds at their disposal are limited, and, unaided, they would have been unable to have carried on excavations except upon a very moderate scale. Owing to the munificence of Sir Thomas Elder, the search party, under charge of Mr. Hurst, which has succeeded in sending down a sufficient number of specimens to whet the appetite of scientists, and even of the curious public, who have no special knowledge of palæontology, for more, will now be reinforced by Dr. Stirling, Mr. Zietz, and another member of the Museum staff. More than that, provision will now be made for the conveyance to Adelaide of larger portions of the remains than it has hitherto been possible to despatch.

"Mr. Hurst has laboured under great disadvantages in the past, owing to the lack of facilities for transporting to the city the results of his excavations. It is to be borne in mind that the fossils are spread over a large tract of country embracing several square miles in extent. Moreover, this singular burial-place of the marsupial and other monsters, which were wont to roam over the continent of Australia, is situated in a remote and comparatively waterless region, so that the difficulties in the way of disinterment and removal are exceedingly great. It is only fitting that those officials of the Museum, who are most skilled in natural history, should be on the spot to superintend operations, and we may rest satisfied that everything that can be done to secure the best illustrations obtainable of the fossil remains will be done. The object to be aimed at is to procure complete specimens, and, from what has been said by those who have visited the locality, it should be practicable to achieve this. Obviously no pains or expense should be spared to bring together, bone to his bone, the framework of the gigantic creatures belonging to a bygone age, whose relics are embedded in the hardened mud and *débris* of what were once the huge swamps of the Far North-East."

Professor Newton, of Cambridge, also received a letter from Dr. Stirling, which was published in *The Proceedings of the Zoological Society* for May 2, 1893. From this letter we learn that *Diprotodon* has five well-developed toes; the bones of the marsupium, or pouch, have been found, and several distinct impressions of the skin of the fore foot were obtained. The locality is six hundred miles north of Adelaide, and, during the dry season, all the carting and travelling has to be done by camels, with a temperature of 111° Fahr. in the shade!—to say nothing of myriads of flies and frequent sandstorms.

With regard to the geological formation of this salt-lake district of South Central Australia, Lake Mulligan is a vast level expanse of salt-encrusted, black mud, only becoming filled after very heavy rains. It is about eight miles wide, and the bones are found somewhere about midway between the east and west edges. Usually the salt crust is not firm enough for bullock traffic, and it is probable that thousands of bullocks have at different times been bogged in crossing or attempting to cross.

Several skeletons of a large wombat, about the size of a bullock, have been unearthed, probably *Phascolomys gigas;* also another slender creature about the size of a sheep, which is as yet an unknown animal. The meaning of this great find of bones would appear to be that immense herds of *Diprotodon* and other creatures, in seeking for water in a dry season, got bogged, just as cattle do now in the North, by hundreds. Crocodiles and alligators inhabited the fresh-water lakes that once covered parts of the district, either in the Pliocene or Pleistocene period—or both.

Probably in a few months' time, drawings and photographs of the skeleton will reach this country. Mr. Smit's drawing, Fig. 75, must only be regarded as a provisional restoration.

CHAPTER XII.

RELICS OF THE PLEISTOCENE AGE.

"Palæontology further teaches that not only the individual, but the species, perishes; that as death is balanced by generation, so extinction has been concomitant with the creative power which has produced a succession of species; and furthermore, that in this succession there has been 'an advance and progress in the main.'"—SIR RICHARD OWEN.

So numerous are the various forms of life which flourished during the Pleistocene period of the world's history, that a separate volume might well be devoted to their consideration. It will only be possible, within the limits of our present concluding chapter, to select one or two prominent examples for description. Those we have chosen are all, with one exception, creatures which once inhabited the continent of South America. The country which has afforded most information with regard to the extinct fauna of this continent is the Argentine Republic, which includes not only Buenos Ayres and the adjacent provinces forming Argentina proper, but likewise the whole of Patagonia. Here is a vast tract of country forming one boundless level plain made by an alluvial deposit of rich black mud, brought down from the higher lands of the interior by the tributaries of the Rio de la Plata, and which constitutes the most extensive pasture-land in the world.

This pampean formation is of fresh-water origin, and was probably formed to a large extent in marshes and swamps. There is, however, evidence that portions of the deposit have been submerged beneath the sea. It is now known to be the tomb of thousands of extinct animals, and their bones tell us that the country was once inhabited by a wonderfully strange fauna, such as has not been found in any other part of the world. At times they are found sticking out from the perpendicular cliffs bordering the river-valleys, while many are met with in sinking wells or making excavations.

Entire skeletons have been obtained during the work of excavating the docks at La Plata or Buenos Ayres. Mr. R. Lydekker, an eminent palæontologist, has lately paid a visit to South America in order to report on the fossil treasures contained in the museums of Buenos Ayres and La Plata. Speaking of the latter, he says, "There the visitor will be absolutely lost in astonishment at the long array of perfect mounted skeletons of numbers of these creatures, while the unmounted skeletons and isolated bones displayed in the wall-cases will convince him that I am not exaggerating when I speak of Argentina as a land of skeletons."[1]

Thanks to the efforts of the celebrated palæontologist, Dr. Moreno, the enchanted city of La Plata now possesses a magnificent museum worthy of a country so richly endowed with fossil treasures. Dr. Moreno, the director of this museum, has made it his principal aim to illustrate the whole fauna, both recent and fossil, of the Argentine Republic. Side by side with the huge skeletons of forms of life which have perished, he has placed for comparison skeletons of the living animals to which they are most nearly related, and it would be well if all collections of fossils were illustrated in the same happy manner.

The *Toxodon* is one of those extinct animals with very "mixed" relationship: a "generalised" hoofed animal, from the Pleistocene deposits of South America, presenting characters which *seem* to connect it with the odd-toed ungulates (rhinoceros, horse, tapir),

[1] The student will do well to consult Mr. Lydekker's valuable papers in *Natural Science* for January and February, 1894, as well as in *Knowledge* for January, February, and March, 1894.

and at the same time with such different groups as the elephants (*Proboscidia*) and the gnawing mammals (rodents). It takes its name from the fact that the teeth are curved like a bow.[1] The skull (see Fig. 76), which looks at first sight rather like that of a horse, may be seen in the Natural History Museum (Gallery No. 1, Pier-case 20). *Toxodon* was discovered years ago by Charles Darwin in the following manner. During his sojourn in Banda Oriental, having heard of some "giant's bones" at a farmhouse on the Sarandis, a small stream entering the Rio Negro, he rode there,

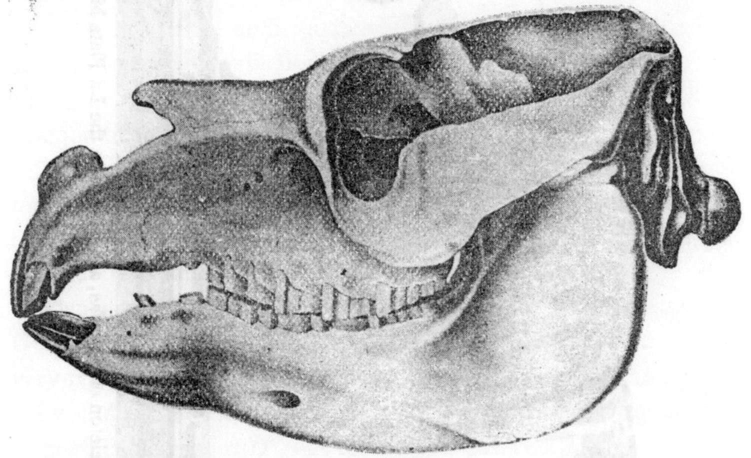

FIG. 76.—Skull of *Toxodon*, from Pleistocene strata. (After Burmeister.)

and purchased for the sum of eighteenpence the skull which has been described by Sir R. Owen. The people at the farmhouse told Mr. Darwin that the remains were exposed by a flood having washed down part of a bank of earth. When found, the head was quite perfect, but the boys knocked the teeth out with stones, and then set up the head as a mark to throw at. Mr. Darwin found a perfect tooth, which exactly fits one of the sockets of this skull, embedded by itself in the banks of the Rio Tercero,

[1] Greek—*toxon*, bow ; *odous*, *odontos*, tooth.

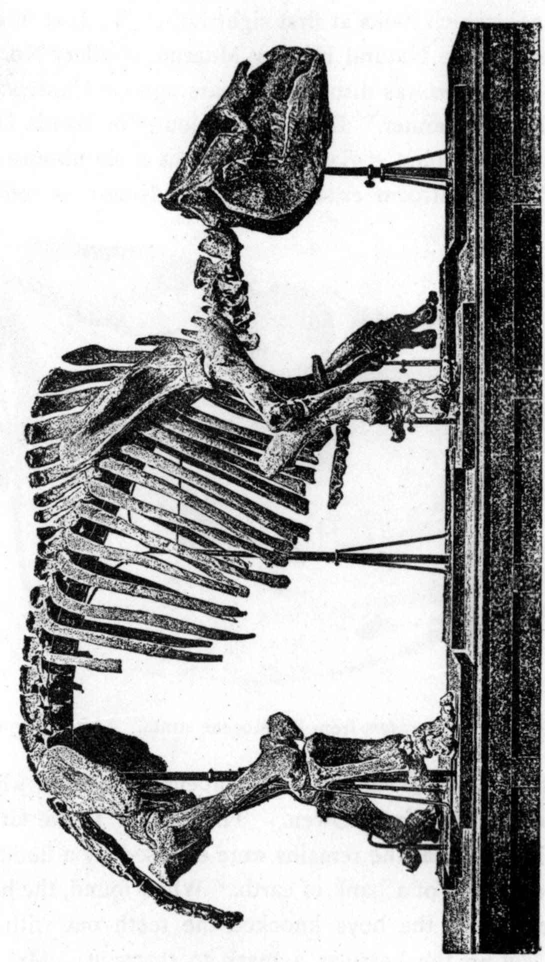

FIG. 77.—Skeleton of *Toxodon*, as mounted in the La Plata Museum.

at the distance of about a hundred and eighty miles from the farmhouse. He found remains of this extraordinary animal at two other places; so that it probably was common at one time. The skull is two feet four inches long.

Speaking of the fossils of this region, he says, "The number of the remains embedded in the grand estuary deposit which forms the Pampas and covers the granitic rocks of Banda Oriental, must be extraordinarily great. I believe a straight line drawn in any direction through the Pampas would cut through some skeleton or bones. Besides those which I found during my short excursions, I heard of many others, and the origin of such names as "the stream of the animal," "the hill of the giant," is obvious. At other times I heard of the marvellous property of certain rivers, which had the power of changing small bones into large; or, as some maintained, the bones grew. As far as I am aware, not one of these animals perished, as was formerly supposed, in the marshes or muddy river-beds of the present land, but their bones have been exposed by the streams intersecting the subaqueous deposit in which they were originally embedded. We may conclude that the whole area of the Pampas is one wide sepulchre of these extinct quadrupeds." Darwin also mentions that he found in the pampean formation the remains of nine great quadrupeds, and many detached bones embedded on the beach near Buenos Ayres, all within the space of two hundred yards square.

But to return to our *Toxodon.* Two complete skeletons of this strange monster have been set up in the La Plata Museum. A skull was described by Owen in the year 1840, who pointed out some supposed resemblances to the Sirenia (Dugong and Manatee). Fig. 77 shows one of the complete skeletons in the La Plata Museum.[1]

[1] We are indebted for this illustration to the courtesy of the proprietors of *Natural Science.* A forthcoming issue of the *Annals* of the La Plata Museum will contain an account of Mr. Lydekker's work there.

The sub-order Toxodontia was founded by Owen for this creature and some of its allies to be mentioned presently. There are ten species of *Toxodon*, of which *T. platensis* is the best known. Burmeister, Ameghino, and others concluded that the Toxodonts were chiefly herbivorous, and lived in marshy plains. The one figured here was nearly as large as an Indian rhinoceros. Cope pointed out certain resemblances to the elephants (Proboscidia).

A peculiarity which the reader will notice at once is the difference in size between the fore and the hind limbs. It was very low in the fore-quarters, in consequence of which its enormous head was carried much below the line of the back. We have not ventured to attempt a restoration of this most puzzling extinct form of life—for one reason, because there was not time to do so before going to press; and secondly, on account of the great peculiarities it presents. In time, perhaps, our leading authorities may be able to form a better idea of what the creature was like when alive. In some respects it may have had a certain resemblance to a rhinoceros, as shown by its relatively short and stout neck and limbs. While the number of toes in each limb is three, its skull shows a curious mixture of characters presented by both the odd-toed and the even-toed hoofed animals. Its teeth were very peculiar; the upper jaw (see Fig. 76) was provided with two pairs of growing chisel-teeth, like the single pair seen in the jaw of a hare or a rabbit, and in all the rodents; and in the lower jaw were three pairs of nearly similar teeth placed horizontally. Only in the little *Hyrax*, or coney, among living animals do we see such permanently growing incisor teeth. The molar teeth likewise grew throughout life instead of forming roots, but the grinding surfaces of these teeth are not at all similar to those of hoofed animals of the present day.

Allied to the *Toxodon* was another and smaller form to which Owen gave the name *Nesodon*, of which the first fragmentary remains were brought to Europe by Darwin in the *Beagle*.

The name of the genus is derived from a well-marked island-like lobe on the inner side of upper molars. This genus is represented in the La Plata Museum by a vast series of remains, including many perfect skulls, as well as jaws, teeth, and limb-bones. Mr. Lydekker recognises only three species, of which the smallest was not much larger than a sheep. He has been fortunate enough to identify some of the limb-bones, to which little attention had been paid by others, and thus to confirm the conclusion that *Nesodon* and *Toxodon* were closely related. The Nesodons, in the structure of their molar teeth, show an approach to the odd-toed ungulates,—the horse, rhinoceros, and tapir: although growing for a considerable time, these eventually formed roots in the ordinary manner. The general structure of the skeleton is in many respects much less "specialised" than is that of the *Toxodon.* A larger species was the *Nesodon imbricatus*, which approached the dimensions of a rhinoceros. There is no reason to think that the one genus was descended from the other. Both must have branched off from some primitive ancestral type of ungulate, and then found their way into South America. An upper jaw and a lower jaw of *Nesodon ovinus* from the south-west coast of Patagonia may be seen in the Natural History Museum; they were brought home by Admiral Sir R. J. Sulivan, K.C.B.

In another highly interesting member of the same sub-order, namely, the little *Typotherium*, the rodent-like character of the teeth is still more striking. This creature probably had claws on its feet instead of hoofs, and may have resembled in appearance the existing capybara. Another form, the *Pachyrucus*, from the deposits of Monte Hermoso, was about the size of a rabbit, and so like a rodent in its general structure that it might easily be mistaken for a near relation of the hares! But it clearly belongs, in spite of these artificial resemblances, to the Toxodont group; this curious phenomenon, of totally distinct groups mimicking each other, is familiar to palæontologists, who call it "parallelism."

Mr. Lydekker proposes the name Astrapotheria for another

group of extinct South American ungulates, including the genus *Homalodontotherium* (first described by Sir William Flower in the *Philosophical Transactions* for 1874), and others. Unlike the Toxodonts, they had their teeth rooted at an early age, and they were of considerable size. Until recently the genus *Homalodontotherium* was only known by specimens of teeth and jaws, but the La Plata Museum contains numerous specimens of vertebræ and limb-bones.

In the gigantic *Astrapotherium* (the type species of which was originally described by Owen as *Nesodon magnum*) there are not so many teeth, and each jaw had a huge pair of tusks, while the upper molars are extremely like those of a rhinoceros—another curious case of parallelism. The presence of these great sharp tusks has suggested to some authorities a relationship with the Dinocerata described in our former work, but this is quite out of the question. Two other highly interesting allied forms are the *Trigodon* (*Toxodontotherium*) and the *Zotodon*, represented by skulls and jaws in the La Plata Museum.

Perhaps the strangest of all the peculiar ungulates discovered in recent Tertiary deposits in South America, is the *Macrauchenia*, so named on account of its long neck.[1]

Burmeister has published a beautiful lithograph of the whole skeleton,[2] of which Fig. 78 is a reduced drawing. Originally supposed to have been allied to the llama, though much larger, it is now known to be a highly specialised form of ungulate, its true affinities being still undecided. Some of its bones brought from South Patagonia by Darwin, and described by Owen, may be seen in the Natural History Museum (Gallery 1, Pier-case 8). Darwin gives the following account of his discovery: "At Port St. Julian, in some red mud capping the gravel on the ninety-feet plain, I found half the skeleton of the *Macrauchenia patachonica*, a remarkable quadruped full as large

[1] Greek—*makros*, long; *auchen*, neck.

[2] *Annales del Museo Publico Buenos Aires.* H. Burmeister, M.D.

as a camel. It belongs to the same division of the Pachydermata with the rhinoceros, tapir, and palæotherium; but in the structure of the bones of its long neck it shows a clear relation to the camel, or rather to the guanaco and llama. From recent sea-shells being found on two of the higher step-formed plains, which must have been modelled and upraised before the mud was deposited in which the *Macrauchenia* was entombed, it is certain that this curious quadruped lived long after the sea was inhabited by its present shells. I was at first much surprised how a large quadruped could so lately have subsisted in latitude 49′ 15″ on

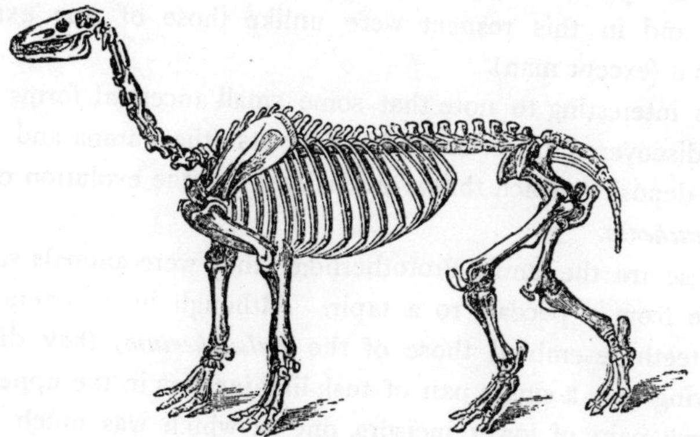

FIG. 78.—Skeleton of *Macrauchenia*, from Pleistocene strata, Buenos Ayres. (After Burmeister.)

these wretched gravel plains with their stunted vegetation; but the relationship of the *Macrauchenia* to the guanaco, now an inhabitant of the most sterile parts, partly explains the difficulty."

Darwin, however, was mistaken in supposing that *Macrauchenia* was related to the llamas and camels, in spite of the similarity in the vertebræ of the neck. Its limbs remind us of some of the fossil horses described in Chapter X.; the teeth are more like those of a rhinoceros. A most peculiar feature of this animal is the way in which the nasal openings are placed on the top of the skull, almost between the eyes. In most mammals the aperture of the nose is quite at the forward end of the skull. Among living animals the elephant and the dugong and the whales have this opening placed far back. What was the reason for such a backward position in the case of the *Macrauchenia* we cannot tell; but we find it difficult to believe with Mr. Lydekker that its head was provided with a trunk—even a short one. Dr. Burmeister has ventured to give it a long trunk; but surely such an appendage was quite unnecessary, considering that with such a long neck its head could easily reach the ground? The jaws were provided with an even and uninterrupted series of teeth, and in this respect were unlike those of any existing mammal (except man).

It is interesting to note that some small ancestral forms have been discovered in certain older deposits (the Parana and Patagonia deposits) which throw much light on the evolution of the *Macrauchenia.*

These are the family Prototheriidæ; they were animals varying in size from a peccari to a tapir. Although in a general way their teeth resembled those of the *Palæotherium*, they differed in having only a single pair of tusk-like incisors in the upper jaw, and two pairs of lower incisors, one of which was much larger than the other.

But the most interesting point about these creatures was the structure of their feet; for we find many of them with feet almost exactly like the European *Hipparion* described in Chapter X., which the reader will remember had one big toe, with two little ones one on each side of it, and it is probable that in one genus, at least, there was only one toe, as in a modern horse; and yet they were *not* horses! Professor Cope regards them as a distinct sub-order—the Litopterna. They differ from horses in having long vertebræ in the neck like those of *Macrauchenia*, which are camel-like, but have flat sides to the bodies (or centra), instead of a ball-and-socket joint. The strata in which these interesting remains occur cannot be nearly so old as our European Eocene or Miocene. It is very interesting to find that while the evolution of the true horses was taking place in Europe and North America, a race of creatures not at all allied to them, but related to the *Macrauchenia*, were developing the same kinds of feet in order to acquire running powers, and thus as it were imitating them. This must be another case of "parallelism."

The extinct ungulates of Pleistocene age in South America were apparently not left unmolested, for the remains of carnivorous animals have been found in the same strata. One very formidable enemy of the Macrauchenia and its allies must have been the terrible sabre-toothed tiger, *Machairodus necator*, also known as *Smilodon* (see Fig. 79); its great curved and serrated

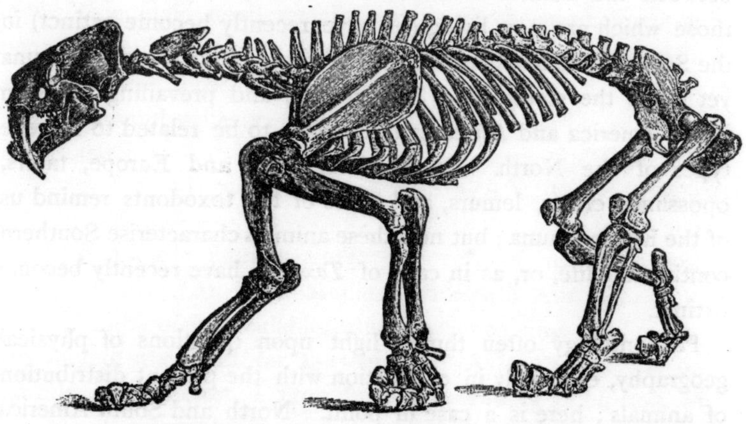

FIG. 79.—Skeleton of *Machairodus* (*Smilodon*), Pleistocene. (After Burmeister.) [The original drawing was made from a photograph; consequently the fore limbs, being near the camera, appear too large in proportion to the hind limbs.]

tusks were actually twelve inches long and eight inches beyond the gums! They could not have been used in the ordinary way, because the creature was unable to open its mouth wide enough; consequently anatomists have been greatly puzzled to know how they were used. Some say that they were of no use, and that it is a case in which "over-specialisation" led to extinction. In our restoration, seen in Plate XXIII., where the Macrauchenia is represented as being attacked by this ferocious animal, a possible use is suggested, namely, that they might help in enabling the creature to hang on to its prey. Strange as it may seem, there is very little difference between the skull of the lion and that of the tiger, and some persons may be inclined to consider *Machairodus* to have been a lion; but most authorities think it was a tiger. In size it must have been equal to the largest lion, and doubtless was equally ferocious, or even more so.

Palæontologists have discovered a very interesting connection between the extinct mammalian animals of the Northern and those which are now living (or have recently become extinct) in the Southern hemisphere. Much of the present Australian fauna yet bears the stamp of a Jurassic age, and prevailing types in South America and Africa can be shown to be related to Eocene types of the North. In North America and Europe, tapirs, opossums, civets, lemurs, and allies of the toxodonts remind us of the Eocene fauna; but now these animals characterise Southern continental life, or, as in case of *Toxodon*, have recently become extinct.

Palæontology often throws light upon questions of physical geography, especially in connection with the present distribution of animals; here is a case in point. North and South America now constitute two distinct zoological provinces, and the table-land of Mexico, by forming a barrier, presents an obstacle to the migration of species which few have been able to surmount. South America is characterised by possessing many peculiar gnawers, a certain family of monkeys, the llama, peccari, tapir, opossums, and especially several genera of edentata, the order which includes the sloths, ant-eaters, and armadilloes. Some

MACRAUCHENIA AND THE SABRE-TOOTHED TIGER (MACHÆRODUS).

From the Pampas formation.

PLATE XXIII.

the Northern country. Now North America has some peculiar rodents, four hollow-horned ruminants, such as the ox, sheep, antelope, of which great group the Southern continent has not one that is indigenous. But a very short time ago geologically their respective faunas were far more alike.

"The more I reflect on this case," says Darwin, "the more interesting it appears: I know of no other instance where we can almost mark the period and the manner of the splitting up of one great region into two well-characterised zoological provinces. The geologist who is fully impressed with the vast oscillations of level which have affected the earth's crust within late periods, will not fear to speculate on the recent elevation of the Mexican platform, or, more probably, on the recent submergence of land in the West Indian archipelago, as the cause of the present zoological separation of North and South America. The South American character of the West Indian mammals seems to indicate that this archipelago was formerly united to the Southern continent, and that it has subsequently been an area of subsidence."

At the present day the pampas region of South America is almost destitute of trees, and may be described as a boundless sea of grass (pampas grass), with only a few slender herbs and trefoils which maintain a frail existence among the tall tussocks; but the strong grass crowds out most plants. Here and there are patches where the grass does not grow, and these are carpeted by small creeping herbs of a livelier green, and are gay in spring with flowers.

This vast tract of country was formerly inhabited by the guanaco, the pampas deer, the viscacha, and the rhea, which were almost the only animals of any size to be found in the whole region. Since then, horses, cattle, and sheep have been introduced by man, and these now roam over the whole district in vast herds. Great changes have taken place here, both in the animal and vegetable kingdoms. Contrasting the former with the present state of South America, Darwin says, "It is impossible to reflect on the changed state of the American continent without the deepest astonishment. Formerly it must have swarmed with great monsters; now we find mere pigmies, compared with the antecedent allied races. If Buffon had known of the gigantic sloth and armadillo-like animals, and of the lost pachydermata, he might have said with a greater semblance of truth that the creative force in America had lost its power, rather than that it had never possessed great vigour. The greater number, if not all, of these extinct quadrupeds lived at a late period, and were the contemporaries of most of the existing sea-shells. Since they lived, no very great change in the form of the land can have taken place. What, then, has exterminated so many species and whole genera? The mind at first is irresistibly hurried into the belief of some great catastrophe; but thus to destroy animals, both large and small, in Southern Patagonia, in Brazil, on the Cordillera of Peru, in North America, up to Behring's Straits, we must shake the entire framework of the globe. An examination, moreover, of the geology of La Plata and Patagonia leads to the belief that all the features of the land result from slow and gradual changes. It appears from the character of the fossils in Europe, Asia, Australia, and in North and South America, that those conditions which favour the life of the larger quadrupeds were lately coextensive with the world: what those conditions were no one has yet even conjectured."

How is it, then, that the forests have disappeared, and all the large mammals have become extinct? The palæontologist is obliged to confess that the causes of extinction (apart from the action of man as an exterminator) offer a very difficult problem, and one which is at present by no means solved. Death in the individual, though a mysterious event, is one with which we are familiar; but the death of a species, a genus, a family, or a whole order (*e.g.* the Dinosaurs), constitutes a mystery on which at present we can only speculate somewhat vaguely. Is there a definite limit to life imposed at first? or is this limit left, to be determined by circumstances? Perhaps both; but if the latter, what are those circumstances?

Sir R. Owen has pointed out that large land-animals are, on account of their size, more liable to extinction, while smaller ones can escape. Any change in the surrounding circumstances is more likely to tell against a large animal than against a small one. Thus, in a long dry season the large mammal will suffer sooner from drought than a small one. Or if, by some change in climate, the available quantity of vegetable food should be greatly diminished, the more bulky creatures will be the first to feel the effects of a stinted diet. Mere bodily strength does not always give victory in the battle of life. Before man inhabited India or Africa, some cause must have checked the continued increase of the elephant. The late Dr. Falconer thought it was chiefly insects, which from incessantly harassing and weakening the elephant in India, check its increase. Certain it is that insects and blood-sucking bats determine the existence of the larger naturalised quadrupeds in several parts of South America. Again, if, by some geographical change, new creatures from a country formerly separated manage to find their way in, they will begin at once to compete with those already there, and in this competition the larger animals, not being able to hide themselves, may be at some disadvantage. Lastly, small quadrupeds are more prolific than large ones, and consequently large numbers may die without any danger of the race becoming extinct.

It is like the old fable of "the oak and the reed;" for the smaller forms can bend and accommodate themselves to changes which are fatal to their more ponderous contemporaries.

"We do not steadily bear in mind," says Darwin,[1] "how profoundly ignorant we are of the condition of existence of every animal; nor do we always remember that some check is constantly preventing the too rapid increase of every organised being left in

[1] *Naturalist's Voyage*, ch. viii.

a state of nature. The supply of food, on an average, remains constant; yet the tendency in every animal to increase by propagation is geometrical, and its surprising effects have nowhere been more astonishingly shown than in the case of European animals run wild during the last few centuries in America. Every animal in a state of nature regularly breeds; yet, in a species long established, any *great* increase in numbers is obviously impossible, and must be checked by some means. We are nevertheless seldom able with certainty to tell in any given species at what period of life, or at what period of the year, or whether only at long intervals, the check falls; or, again, what is the precise nature of the check. Hence, probably, it is that we feel so little surprise at one, of two species closely allied in habits, becoming rare and the other abundant in the same district; or, again, that one should be abundant in one district, and another, filling the same place in the economy of nature, should be abundant in a neighbouring district, differing very little in its conditions. If asked how this is, one immediately replies that it is determined by some slight difference in climate, food, or the number of enemies; yet how rarely, if ever, we can point out the precise cause and manner of action of the check! We are, therefore, driven to the conclusion that causes generally quite inappreciable to us, determine whether a given species shall be abundant or scanty in numbers. . . . Who could feel any great surprise at hearing that the Megalonyx was formerly rare compared with the Megatherium, or that one of the fossil monkeys was few in number compared with one of the now living monkeys? and yet, in this comparative rarity, we should have the plainest evidence of less favourable conditions for their existence. To admit that species generally become rare before they become extinct—to feel no surprise at the comparative rarity of one species from another, and yet to call in some extraordinary agent and to marvel greatly when a species ceases to exist—appears to me much the same as to admit that sickness

THE URUS OF CÆSAR (BOS PRIMIGENIUS).

PLATE XXIV.

in the individual is the prelude of death; to feel no surprise at sickness, but when the sick man dies to wonder, and to believe that he died through violence."

When the Roman armies first penetrated the dark forests of Germany and Belgium, there were found, among other beasts of the chase, two large species of wild oxen; the one called "Bison" distinguished by its shaggy coat, the other called "Urus" by the great size of its horns. Both were occasionally captured and exhibited alive in the shows of the amphitheatre. Seneca thus briefly defines the characters of the two species—

> "*Tibi dant variæ pectora tigres,*
> *Tibi villosi terga bisontes,*
> *Latisque feri cornibus uri.*"

Pliny also distinguishes the one from the other. The great Bison still survives, by virtue of strict protective laws, in Lithuania, and also in the Caucasus. Fine specimens from both localities may be seen in the Gallery of Stuffed Mammals in the Natural History Museum. It is distinguished from all the domestic cattle of Europe by the thicker clothing of hair, which, in the male Bisons, is developed into a curly mane; a feature noticed by the Romans. The Bison differs from the Ox in several particulars. That the former existed in this country is abundantly proved by the skulls and horn-cores discovered in the latest geological deposits in various parts of Great Britain. Specimens may be seen in the Natural History Museum (Gallery 1, Pier-case 16), obtained from the "brick-earth" of Ilford and Walton in Essex; from Erith, Kent; Pecham, Surrey; from Wiltshire and Lincolnshire. It had a wide geographical range. A specimen from Worcestershire measures three feet four inches across the horn-cores from tip to tip. The species must have been contemporary with the mammoth, woolly rhinoceros, and the hyæna.

The *Bos primigenius*, or Urus of Cæsar, has been found in the same localities. A magnificent specimen, with skull and horn-cores, is preserved in the Natural History Museum (Gallery 1, Pier-case 18). It was found at Blair Athol, Perthshire. A similar large skull (*Bos longifrons ?*), now in the Museum of the Royal Irish Academy, has a cut in the forehead into which can be accurately fitted several of the narrow bronze "celts" or arrow-heads, so frequently dug up in Ireland, which may perhaps be taken to prove that it was killed by the hunter's arrow. The most complete series of bones of one individual specimen of the *Urus* was obtained from the drift overlying the London clay at Herne Bay. The length of each horn-core along the outer curve is thirty-nine inches, and the circumference of the core at its base over eighteen inches. According to Mr. Woods, the skull and horns of the great Urus were discovered in a tumulus on the Wiltshire Downs. Many bones of deer, boars, etc., were discovered at the same time; also several fragments of pottery of ancient British manufacture. Mr. Woods suggests that the conflicts of the first settlers in Britain with the Uri might have given rise to the legends of Guy, Earl of Warwick, killing the Dun Cow.

Cæsar, describing under the name Urus certain wild oxen of the great Hergynian forest, says, "These Uri are little inferior to elephants in size, but are bulls in their colour, nature, and figure. Great is their strength, and great their swiftness; nor do they spare man or beast when once they have caught sight of him. These, when trapped in pitfalls, the hunters unsparingly kill. The youths, exercising themselves by this sort of hunting, are hardened by the toil; and those among them who have killed most, bringing with them the horns, as testimonials, acquire great praise. But these Uri cannot be habituated to man, nor made tractable, not even when taken young. The great size of the horns, as well as the form and quality of them, differs much from our oxen."

It has been conjectured by Cuvier and other naturalists that our domestic cattle are the degenerate descendants of the great

Urus. The latter was probably the ancestor of the larger existing European varieties found in Spain, Italy, and Hungary. The wild cattle at Chillingham Park, Northumberland, may perhaps be the last surviving descendants of the *Bos primigenius*, but much degenerated in size. According to Flower and Lydekker,[1] the existing ox (*Bos taurus*) is only a small variety of the Urus. A smaller race, whose remains have been found in the turbaries and fens of England, and have been described under the names of *B. longifrons* and *B. frontosus*, seems to be only a stunted variety of the same species, from which it is probable that the small cattle of Wales and Scotland have been derived.

In the "Niebelungen Lied" of the twelfth century two kinds of great wild oxen are mentioned, as having been slain in the great hunt of the Forest of Worms, under exactly the same names as those which the Romans gave to them.

> "Dar nach schluch er schiere, einem Wisent und einem Elch
> Starcher Ure vier, und einem grimmen Schelch."
>
> "After this he straightway slew a Bison and an Elk,
> Of the strong Uri four, and a single fierce Schelch."

A later poem says—

> "Mightiest of all the beasts of chase
> That roam in woody Caledon,
> Crushing the forest in his race,
> The Mountain Bull comes thundering on."

[1] *Mammals*, p. 367.

APPENDIX I.

TABLE OF STRATIFIED ROCKS.

Periods.		Systems.	Formations.	
CAINOZOIC.	Quaternary.	**RECENT**	Terrestrial, Alluvial, Estuarine, and Marine Beds of Historic, Iron, Bronze, and Neolithic Ages	Dominant type, Man.
		PLEISTOCENE (250 ft.)	Peat, Alluvium, Loess Valley Gravels, Brickearths Cave-deposits Raised Beaches Palæolithic Age Boulder Clay and Gravels	
	Tertiary.	**PLIOCENE** (100 ft.)	Norfolk Forest-bed Series Norwich and Red Crags Coralline Crag (Diestian)	Dominant types, Birds and Mammals.
		MIOCENE (125 ft.)	Œningen Beds Freshwater, etc.	
		EOCENE (2600 ft.)	Fluvio-marine Series (Oligocene) Bagshot Beds London Tertiaries } (Nummulitic Beds)	
SECONDARY, OR MESOZOIC.		**CRETACEOUS** (7000 ft.)	Maestricht Beds Chalk Upper Greensand Gault	
		NEOCOMIAN	Lower Greensand Wealden	
		JURASSIC (3000 ft.)	Purbeck Beds Portland Beds Kimmeridge Clay (Solenhofen Beds) Corallian Beds Oxford Clay Great Oolite Series Inferior Oolite Series Lias	Dominant type, Reptilia.
		TRIASSIC (3000 ft.)	Rhætic Beds Keuper Muschelkalk Bunter	

TABLE OF STRATIFIED ROCKS—*Continued.*

Periods.	Systems.	Formations.	
PRIMARY, OR PALÆOZOIC.	**PERMIAN** or **DYAS** (500 to 3000 ft.)	Red Sandstone, Marl } Zechstein Magnesian Limestone, etc. } Red Sandstone and Conglomerate Rothliegende	
	CARBONIFEROUS (12,000 ft.)	Coal Measures and Millstone Grit Carboniferous Limestone Series	Dominant type, Fishes.
	DEVONIAN & OLD RED SANDSTONE (5000 to 10,000 ft.)	Upper Old Red Sandstone Devonian Lower Old Red Sandstone	
	SILURIAN (3000 to 5000 ft.)	Ludlow Series Wenlock Series Llandovery Series May Hill Series	
	ORDOVICIAN (5000 to 8000 ft.)	Bala and Caradoc Series Llandeilo Series Llanvirn Series Arenig and Skiddaw Series	
	CAMBRIAN (20,000 to 30,000 ft.)	Tremadoc Slates Lingula Flags Menevian Series Harlech and Longmynd Series	Dominant type, Invertebrata.
	EOZOIC—ARCHÆAN (30,000 ft).	Pebidian, Arvonian, and Dimetian Huronian and Laurentian	

APPENDIX II.

CLASSIFICATION of Fishes, Amphibians, Reptiles, and Mammals, including extinct types, *which are in italics.* The scheme here given is partly taken from Nicholson and Lydekker's *Manual of Palæontology*, third edition, 1889, and may be of some use to students.

VERTEBRATE ANIMALS.

Sub-Kingdom : VERTEBRATA.

CLASS I. **Fishes** (PISCES).

ORDER I. **Cyclostomi.** *Ex.* Hag-fishes, Lampreys. This order is probably unrepresented in the geological record.

ORDER II. **Elasmobranchei.** *Ex.* Sharks (Carcaridæ), Dog-fishes (Scylliadæ), Saw-fishes (Pristis), Rays and Skates (Raiidæ), *Hybodus* (Jurassic).

ORDER III. **Chimæroidei** (or **Holocephala**). *Ex.* Chimæra, *Edaphodon.* The Chimæroids are regarded by some writers as a sub-order of the Elasmobranchei.

ORDER IV. **Dipnoi.** *Ex.* African Mud-fish (Protopterus), the Australian Mud-fish (Ceratodus), the South American Mud-fish (Lepidosiren), *Dipterus* (Devonian Age). Dr. Gunther regards this order as a subdivision of the Ganoids. They are fresh-water fishes.

ORDER V. **Ganoidei.** *Ex.* Sturgeon (Acipenser), Gar pike of North America (Lepidosteus), Bony Pike of the Nile (Polypterus), *Holoptychius* (Old Red Sandstone), *Rhizodus* (Carboniferous), *Undina* (Jurassic). There is considerable diversity of opinion with regard to the classification of

Ganoids, but the system of Dr. Traquair is followed here. The remarkable families represented by the Old Red Sandstone form, *Cephalaspis*, *Pterichthys*, *Coccosteus*, *Pteraspis*, *Bothriolepis*, etc., are here included in the Ganoids, but by Smith, Woodward, and others are considered to be so different as to require a place for themselves—perhaps more than one distinct order.

ORDER VI. **Teleostei (or Bony Fishes).** *Ex.* Eels (Murænidæ), Salmon and Trout (Salmonidæ), *Eurypholis* (Cretaceous).

CLASS II. **Amphibia** (Batrachia of some authors).

ORDER I. *Labyrinthodontia.* Ex. *Labyrinthodon* (*Mastodonsaurus*), *Dolichosoma*, *Keraterpeton*, *Archægosaurus*, *Loxomma*, *Actinodon*, *Dendrerpeton.* Some supposed Amphibia turn out to be *Anomodonts.*

ORDER II. **Apoda.** The Blind-worms (Cæcilia) are unknown in a fossil state.

ORDER III. **Caudata (Tailed Amphibians).** *Ex.* Water-newts (Triton), Salamanders (Salamandra), *Megalobatrachus*, Axolotl (Siredon), Mud-eel (Siren).

ORDER IV. **Ecaudata (Tailless Amphibians).** *Ex.* Frog (Rana), Toad (Bufo).

CLASS III. **Reptilia** (REPTILES).

The provisional arrangement here given is a modification of one recently proposed by Dr. G. Baur, of Newhaven, and may be stated as follows :—

ORDER I. *Anomodontia.* Ex. *Dicynodon*, *Pareiasaurus*, *Oudenodon*, *Galæsaurus*, *Dimetrodon.*

ORDER II. *Sauropterygia.* Ex. *Plesiosaurus*, *Pliosaurus*, *Nothosaurus*, *Polyptychodon.*

ORDER III. **Turtles, Tortoises (Chelonia).** *Ex.* Turtle (Chelone), Tortoise (Testudo).

ORDER IV. *Fish-lizards* (*Ichthyosauria*).

ORDER V. *Proterosauria.* Ex. *Proterosaurus.*

ORDER VI. *Rhynchocephalia.* Ex. Rhynchosaurus, *Hyperodapedon*, Sphenodon.

ORDER VII. **Serpents, Lizards, Chamæleons (Squamata).** *Ex.* Vipers (Viperidæ), Boas and Pythons (Boidæ), *Mosasaurus* (*Mosasauridæ*), *Palæophys* (Eocene).

ORDER VIII. *Dinosauria* (*Ornithoscelida* of Huxley).

SUB-ORDER I. *Ornithopoda* (Bird-footed). Ex. *Hadrosaurus, Iguanodon, Hypsilophodon, Scelidosaurus, Stegosaurus.*

SUB-ORDER II. *Theropoda* (Beast-footed). Ex. *Megalosaurus, Anchisaurus, Compsognathus, Cœlurus.*

SUB-ORDER III. *Sauropoda.* Ex. *Atlantosaurus, Brontosaurus* (*Camarosaurus* of Cope). *Pelorosaurus, Cetiosaurus, Ornithopsis, Hoplosaurus, Diplodochus.*

ORDER IX. **Crocodiles, Alligators, etc. (Crocodilia).**

SUB-ORDER I. *Ætosauria.* Ex. *Ætosaurus, Phytosaurus* (*Belodon*).

SUB-ORDER II. *Parasuchia, Stagonolepis.*

SUB-ORDER III. Eusuchia. Ex. *Teleosaurus, Pelagosaurus, Metriorhynchus, Steneosaurus, Geosaurus, Goniophys, Bernissartia, Diplocynodon,* Crocodilus, Gavialis.

ORDER X. *Pterodactyls* (*Ornithosauria*).

SUB-ORDER I. *Pteranodontia.* Ex. *Pteranodon.*

SUB-ORDER II. *Pterosauria.* Ex. *Pterodactylus, Ornithocheirus, Rhamphorhynchus, Scaphognathus, Dimorphodon.*

CLASS IV. **Birds** (AVES).

ORDER I. *Saururæ.* *Archæopteryx.*

ORDER II. **Struthious Birds (Ratitæ).**

SUB-ORDER I. *Odontoclæ.* Ex. *Hesperornis.*

SUB-ORDER II. *Æpiornithes.* Ex. *Æpiornis.*

SUB-ORDER III. Apteryges. *Ex.* Apteryx.

SUB-ORDER IV. *Immanes.* Ex. *Moa* (*Dinornis*).

SUB-ORDER V. Megistanes. Emeu and Cassowary (Dromæus and Cassuarius), Australia.

SUB-ORDER VI. Rheæ. *Ex.* Rhea (South America).

SUB-ORDER VII. Struthiones. *Ex.* Ostrich (Arabia and Africa).

SUB-ORDER VIII. *Gastornithes.* Ex. *Gastornis, Odontopteryx.*

ORDER III. **Carinatæ** (with keeled breast-bone).

SUB-ORDER I. *Odontormæ.* Ex. *Ichthyornis.*

SUB-ORDER II. Crypturi. *Ex.* Tinamus.

SUB-ORDER III. Penguins, etc. (Impennes).

SUB-ORDER IV. Petrels, etc. Tubinares.

SUB-ORDER V. Grebes, Divers, etc. (Pygopodes). *Ex.* Great Auk (Alca impennis).

SUB-ORDER VI. Gulls and Terns, Gaviæ. *Ex.* Larus (Gull).

SUB-ORDER VII. Woodcock, etc. (Limocolæ). *Ex.* Numenius (Curlew).

SUB-ORDER VIII. Cranes, etc. (Alectores). *Ex.* Grus (Crane).

SUB-ORDER IX. Rails, Coots, etc. (Fulicariæ). *Ex.* Rallus (Rail).

SUB-ORDER X. Cocks, Pheasants, Turkeys, etc. (Gallinæ). Many extinct Tertiary genera.

SUB-ORDER XI. Pigeons, etc. (Columbæ). Ex. *Dodo* (*Didus ineptus*).

SUB-ORDER XII. Geese, etc. (Anseres). *Ex.* Swan (Cygnus).

SUB-ORDER XIII. *Odontopteryges*, *Odontopteryx* (of London Clay).

SUB-ORDER XIV. American Screamers (Palamedeæ, unrepresented in a fossil state).

SUB-ORDER XV. Flamingoes (Odontoglossæ). *Elornis*, of Lower Miocene, probably a Flamingo.

SUB-ORDER XVI. Ibis, Spoonbill, etc. (Herodiones). *Ex.* Ibis pagana, from the Miocene.

SUB-ORDER XVII. Web-footed Birds (Steganopodes). *Ex.* Albatross, etc. *Ex.* Pelican (Pelecanus), of the Miocene.

SUB-ORDER XVIII. Hawks, etc. (Accipitres). *Ex.* American Vulture (Cathartes).

SUB-ORDER XIX. Owls, etc. (Striges). *Ex.* Strix (Owl).

SUB-ORDER XX. Parrots, Cockatoos, etc. (Psittachi). *Ex.* African Parrot (Psittachus).

SUB-ORDER XXI. Kingfishers, Woodpeckers, etc.,—a somewhat heterogeneous group (Picariæ). *Ex.* Halcyornis (a Kingfisher of the London Clay).

SUB-ORDER XXII. Birds of passage (Passares). *Ex.* Crow (Corvus).

CLASS **Mammalia.**

SUB-CLASS I. **Prototheria.**

ORDER I. **Monotremes (Monotremata).** *Ex.* Duck-billed Platypus (Ornithorhynchus paradoxus), the Porcupine Anteater (Echidna aculeata). The imperfectly known *Thylacoleo* may be placed in this Order. Also a number of more ancient forms, such as *Plagiaulax*, *Tritylodon*, *Bolodon*, *Microlestes*.

SUB-CLASS II. **Metatheria.**

ORDER II. **Kangaroos, Opossums, etc. (Marsupialia).**

SUB-CLASS III. **Eutheria.**

ORDER I. **Sloth, Armadillos, etc. (Edentata).** *Ex.* Scaly Ant-eater (Manis), Cape Ant-eater (Orycteropus). Sloth (Bradypus), *Megatherium*, *Glyptodon* (Pleistocene).

ORDER. II. **Whales, Dolphins, etc. (Cetacea).**

SUB-ORDER I. Mystacoceti. *Ex.* Greenland Whale (Balæna).

SUB-ORDER II. *Archæoceti.* EX. *Zeuglodon.*

SUB-ORDER III. Odontoceti. *Ex.* Sperm Whale (Physeter), Dolphin (Delphinus), *Ziphioides* (Miocene), *Squalodon* (Miocene).

ORDER III. **Manatee, Dugong, etc. (Sirenia).** Ex. *Rhytina* (Pleistocene).

ORDER IV. **Hoofed Mammals (Ungulata).**

SUB-ORDER I. Even-toed Ungulates (Artiodactyla). *Ex.* Hippopotamus, Pig (Sus), *Chæropotamus* (Eocene), *Anthracotherium* (Eocene), *Hyopotamus* (Eocene), *Mericopotamus* (Pliocene), *Oreodon* (*Cotylops*), *Anoplotherium* (Eocene), *Cænotherium* (Eocene), *Dichodon* (Eocene), *Tragulus* (Pliocene), *Pöebrotherium* (Miocene), Camel (Camelus), Llama (Auchenia), *Protalbis* (Miocene), *Procamelus* (Pliocene), Stag (Cervus), Giraffe (Giraffa) *Sivatherium*, Antelopes, Goats, Sheep, and Oxen (Bovidæ), *Palæoreas* (Pliocene).

SUB-ORDER II. Odd-toed Ungulates (Perissodactyla). *Ex.* Tapir (Tapirus), *Lophiodon* (Eocene), *Hyracotherium* (Eocene), (*Pliolophus*, *Orohippus*, and perhaps *Eohippus*,) *Pachynolophus*, *Palæotherium* (Eocene), *Anchitherium* (Miocene), *Pliohippus* (Pliocene), Rhinoceros, *Elasmotherium*, *Chalicotherium* (Eocene), *Brontops*, *Titanotherium* (Miocene), *Diplacodon* (Eocene), *Macrauchenia* (Pleistocene).

SUB-ORDER III. *Toxodontia*, *Toxodon* (Pleistocene), *Typotherium* (Pleistocene).

SUB-ORDER IV. *Condylarthra*, *Periptychus* (Eocene), *Phenacodus* (Eocene).

SUB-ORDER V. Hyrachoidea, Cony (Hyrax).

SUB-ORDER VI. *Amblypoda.* Ex. *Pantolamda*, *Coryphodon*, *Dinoceras*, *Uintatherium*, *Tinoceras* (all Eocene).

SUB-ORDER VII. Elephants, etc. (Proboscidia). *Ex.* Elephas, *Mastodon* (Miocene and Pliocene), *Mammoth* (Pleist.), *Dinotherium* (Miocene).

Group *Tillodontia.* Ex. *Tillotherium* (Eocene).

ORDER V. **Hares, Rabbits, etc. (Rodents).**

SUB-ORDER I. Duplicidentata (with two incisors in upper jaw), Hares (Lepus).

SUB-ORDER II. Simplicidentata (one incisor in each jaw). *Ex.* Cavia (Cavy). Porcupine (Histrix), Beaver (Castor), Pouched Rat (Geomys), Rats and Mice (Muridæ), Squirrel (Sciurus).

ORDER VI. **Carnivorous Mammals (Carnivora).**

SUB-ORDER I. Seals, Walrusses, Pinnipedia. *Ex.* Seal (Phoca), Walrus (Trichecus).

SUB-ORDER II. Carnivora vera. *Ex.* Otter (Lutra), Weasel (Mustela), Racoon (Procyon), Bear (Ursus), *Amphicyon* (Eocene), *Ælurodon* (Pliocene), Civet (Viverra), Hyæna, Lion (Felis Leo), *Nimravus* and *Machærodus* (Pleistocene).

SUB-ORDER III. *Creodonta, Hyænodon* (Eocene), *Proviverra* (Eocene), *Apterodon* (Miocene).

ORDER VII. **Insectivorous Mammals (Insectivora).** *Ex.* Mole (Talpa), Hedgehog (Erinaceus). Some fossil forms.

ORDER VIII. **Bats, etc. (Cheiroptera).** *Ex.* Bat (Vespertilio).

ORDER IX. **Lemurs, Monkeys, Apes, Man (Primates).**

SUB-ORDER I. Lemuroidea. *Ex.* Lemur. Some fossil forms.

SUB-ORDER II. Anthropoidea, Gorilla (Troglodytes), Orang (Simia).

Man, Homo sapiens.

APPENDIX III.

LITERATURE.

1. Popular Works.

Works by Doctor Gideon A. Mantell:—
Medals of Creation.
Wonders of Geology.
Petrifactions and their Teaching.
Phases of Animal Life. By R. Lydekker.
Science for All. 5 vols. (Chapters on Extinct Animals.)
Winners in Life's Race. By Arabella Buckley (Mrs. Fisher.)
Romance of Natural History. 2 vols. By H. P. Gosse.
The Autobiography of the Earth. By Rev. H. N. Hutchinson.
Records of the Rocks. By W. S. Symmonds.
The Story of the Earth and Man. By Sir Wm. Dawson.
The Old Red Sandstone. By Hugh Miller.
Sketch-Book of Popular Geology. By Hugh Miller.
Early Man in Britain. By Prof. Boyd Dawkins.
Cave-hunting. By Prof. Boyd Dawkins.
The Horse. By Sir W. H. Flower.
American Addresses. By the Right Hon. T. H. Huxley.
The Living World. W. H. Conn.

2. Works of Reference.

A Manual of Palæontology. 2 vols. By Prof. Alleyne Nicholson and R. Lydekker.

The Life-History of the Earth. By Prof. Alleyne Nicholson.

Origin of Species. By Charles Darwin.

The English Encyclopedia. (The 2 vols. on Natural History contain much information on extinct animals.)

The Encyclopedia Britannica. Eighth or Ninth Editions.

Phillips's *Manual of Geology*. New Edition, by Prof. H. G. Seeley and R. Etheridge.

Prehistoric Europe. By Prof. James Geikie.

Palæontological Memoirs. By Hugh Falconer, M.D.

Mammals Living and Extinct. By Sir William Flower and R. Lydekker.

British Fossil Mammals and Birds. By Sir R. Owen.

A History of British Fossil Reptiles. By Sir R. Owen. 4 vols. (Cassell.) Most of this work has been previously published in the *Monographs of the Palæontographical Society*.

A Manual of Palæontology. By Sir R. Owen.

Memoir on the Archæopteryx. By Sir R. Owen.

Extinct Fossil Mammals of Australia. Sir R. Owen. London, 1887.

Odontography. By Sir R. Owen.

A Guide to the Exhibition Galleries of the Department of Palæontology in the British Museum (Natural History). In two parts. 6*d*. each.

A Catalogue of British Fossil Vertebrata. By A. S. Woodward and C. D. Sherborn.

The Geographical and Geological Distribution of Animals. By Prof. A. Heilprin.

Elements of Geology. By Prof. Joseph Le Conte. New York.

Memoirs of the Geological Survey of Great Britain.

Monographs of the Palæontographical Society.

Palæontographica. Edited by Prof. K. Zittel.

Ichnology of New England. By Edward Hitchcock. Boston, 1858.

Ichnographs from the Sandstone of the Connecticut River. By Dr. J. Deane. Boston, 1861.

Ichnology of Annandale. By Sir Wm. Jardine. Edinburgh, 1853.

Fossil Footsteps in the Red Sandstones of Potsville, Pennsylvania. By Prof. J. Lea.

History and External Characters of the Dodo, Solitaire, and other Extinct Brevipennate Birds of Mauritius, Rodriguez, and Bourbon. H. E. Strickland and A. C. Melville. London, 1848.

A Monograph on the Extinct Toothed Birds of North America. By Prof. O. C. Marsh. New Haven, Connecticut, 1880.

The Vertebrata of the Tertiary Formations. By Prof. E. D. Cope.

The Vertebrata of the Cretaceous Formations of the West. By Prof. E. D. Cope.

Contributions to the Extinct Vertebrata Fauna of the Western Territories. By Joseph Leidy. Washington, 1873.

(The above Monographs by Marsh, Cope, and Leidy are all in the United States Geological Survey of the Territories.)

Cretaceous Reptiles of the United States. Joseph Leidy. Smithsonian Contrib. Knowledge, vol. xiv. (1864).

The Survival of the Fittest. By Prof. E. D. Cope.

Handbuch der Palæontologie. By Prof. K. A. Zittel. Munich, 1888.

Fauna der Gaskohle, und der Kalkstein der Perm-formation Bohemens. By A. Fritsch. Prague, 1883–1886.

Les Enchainements du Monde Animal. By A. Gaudry. Paris, 1883.

Monograph of the Fishes of the Old Red Sandstone of Britain (Cephalaspidæ) Palæontographical Society, 1868–1870. J. Powrie and Ray Lankester.

Die Placodermen. C. H. Pander, 1857.

Geological Survey of Ohio. Palæontology, I., II.

Transactions of the Royal Society of Dublin.

Abhandlungen der Schweizerischen. Paläontologishen Gesellschaft (Batrachia), vol. v.

Reports of the British Association for the Advancement of Science.

A Manual of the Anatomy of Vertebrated Animals. T. H. Huxley. 2nd edit.

3. Journals.

The student will find numerous valuable papers by well-known geologists in the following Journals :—

The Philosophical Transactions of the Royal Society.

The Quarterly Journal of the Geological Society.

The Geological Magazine.

Nature.

Natural Science.

Knowledge. Papers by R. Lydekker.

The American Journal of Science. Papers by Prof. O. C. Marsh.

The American Naturalist. By Prof. E. D. Cope.

Bulletin of the American Museum of Natural History. New York.

Annales del Museo Publico do Buenos Aires. Papers by H. Burmeister.

Annals of the Museum of La Plata.

THE END.